ECO-FRIENDLY BUILDING
CONSTRUCTION FOR REDUCTION OF GLOBAL WARMING

NIRAJAN KUMAR JHA

BlueRose ONE.com
Stories Matter
NewDelhi • London

BLUEROSE PUBLISHERS
India | U.K.

BlueRose ONE.com
Stories Matter
NewDelhi • London

For permissions requests or inquiries regarding this publication,
please contact:

BLUEROSE PUBLISHERS
www.BlueRoseONE.com
info@bluerosepublishers.com
+91 8882 898 898
+4407342408967

ISBN: 978-93-6783-721-4

Cover design: Daksh
Typesetting: Tanya Raj Upadhyay

First Edition: November 2024

DEDICATED TO:

My Late Father Prof.Dr. Brajbhushan Jha and My Late Mother Smt.Indira Devi Jha

My Late Father Prof.Dr.Brajbhushan Jha Receiving "Suprabal Gorkha dakschin Bahu" From His Majesty King Gyanendra Bir Bikram Shah Dev

AUTOBIOGRAPHY OF PROF. DR. BRAJBHUSHAN JHA

Prof.Dr.Brajbhushan Jha was the preeminent leader and Mathematician of the Nepal. Born: January 12th , 1947 , Died: January 26 , 2009, JANAKPUR. Spouse: Indira Devi Jha. Education: S.L.C Board, Nepal (1964), Intermediate Bihar University (1966) , Graduation (Honours) Bihar University (1969), Masters (Mathematics) Tribhuvan University (1971), Ph.D. Patna University (1985) Children: Er.Nabin Kumar Jha(Civil Engineer), Dr.Prabin Kumar Jha (Lecturer, Zoology), Mr.Nutan Kumar Jha (H.A), Er.Nirajan Kumar Jha (Civil Engineer) Occupational Experience: 1. Lecturer, Trichandra College, Kathmandu (1972- 1973) 2. Lecturer, R.R.M Campus, Janakpurdham (1973- 1986) 3. Reader, R.R.M Campus, Janakpurdham (1986-1998) 4. Professor, R.R.M Campus, Janakpurdham (June 5, 1998 -) 5. Chairman, Mathematics Instruction Committee (Sept 6, 1998 -)

PUBLICATIONS:

1. Paper entitled "SOME NEW RESULTS ON DUAL NUCLEAR SPACE " Published in the journal of BIHAR MATHEMATICAL SOCIETY. Vol. 8, 1984, pp 55-58.
2. Paper entitled "SOME NEW RESULTS ON SPECTRAL SUB SPACES OF AUTOMORPHISMS " Published in the journal of BIHAR MATHEMATICAL SOCIETY. Vol.9 1985 pp 1993-1996 .
3. Biographical note Published in Asia/Pacific - Who's Who, Vol. 1, 1998, Pages 231.
4. Biographical Note Published in Asia/American - Who is Who , Volume -1 , 1999 Pages 355 .
5. Biographical Notes Published in Nepal Who is Who 1997 , pages 140.

MEDALS AND AWARDS AWARDED:

1. Awarded, " SUPRABAL GORKHADAKSCHIN BAHU " on the Auspicious Occasion of 56th Birth Anniversary of His Majesty The King Gyanendra Bir Bikram Sah Dev on July 7, 2002
2. Awarded " MAHENDRA VIDYA BHUSHAN GOLD MEDAL " by His Majesty The King Birendra Bir Bikram Shah Dev on February 24, 1992 for securing Ph.D in Mathematics.
3. Awarded " MAHENDRA VIDYA BHUSHAN GOLD MEDAL " by His Majesty The King Birendra Bir Bikram Shah Dev on February 19, 1973 for Securing First-class in M.A (Mathematics).
4. Awarded, the outstanding teacher's reward for doing praiseworthy Job in the field of education as a Teacher on the auspicious occasion of "13th EDUCATION DAY: SHIKHCHA DIVAS " 1988 .
5. Awarded, Tribhuvan University " DIRGHA SEVA PADAK " on July 9 , 2000.

OTHER EXPERIENCES:

1. MEMBER, Tribhuvan University Service Commission Appointed by His Majesty King Gyanendra Bir Bikram Shah Dev on August 5 , 2005.
2. Assistant Evaluation officer Nepal Family Planning and Maternal child health Project on December 5 , 1972 to February 11 , 1973.
3. Warden, R.R.M CAMPUS, Hostel 1975 - 1977.
4. Member, Mathematics Subject Committee on November 13, 1994. to 2008 . Central Department of Mathematics, Institute of Science and Technology , Tribhuvan University, Kathmandu.
5. Member, Vidya Prashad (Faculty Board), Institute of Science and Technology, Tribhuvan University, June 19, 1997 to June 19, 1999.

POLITICAL CONTRIBUTIONS:

1. After National Referendum, 1, First of all, started the third Force as a candidate of Vice President in Nepal University Teacher's Association. Election held in 1986 and secured 540 intellectual votes. Before 1986, only candidates of Democratic and Progressive forces were taking part in that election.
2. Encouraged by the votes obtained by him in the previous election held in 1986, Third Force came in the Nepal University Teacher's Association election held in the form of a panel headed by Late Dr.Chandra Prasad Gorkhali. Dr.Brajbhushan Jha was again included in that panel as a candidate of Vice president.
3. Supported Vigorously PARTYLESS PANCHAYAT System during the critical and momentous period of National Referendum and upheld the cause of the PANCHAYAT SYSTEM in the University.
4. Active participation as a guest speaker deputed by " Panchayat Policy and Evaluation Committee " on the auspicious occasion of " King Mahendra Birth Anniversary and Constitution Day " celebrated at the District headquarters of Sindhuli on December 16, 1986 , the District headquarters of Bara on the December 16, 1987

and the District headquarters of Kanchanpur on the December 16, 1988.

5. Active Participation as a guest Speaker on the auspicious occasion of Birth Anniversary of His Majesty The King Celebrated every year at different Schools as well as colleges and youth organizations of Mahottari District.

My Late Grandfather Shri Bholanath Jha (Former Cabinet Minister and President of W.H.F)

Thinkers have said that living man is only a shadow of death. The death of a born person is certain. However, not all deaths are significant, nor are they the subject of universal interest. Thousands of Nepalis left the world in 2067 B.S. Only some of them came within the interest of the country and society. In the year 2067 B.S, the most important saint leader Krishna Prasad Bhattarai. Bhattarai's death, who was active in Nepali politics for nearly 60 years and synonymous with a simple life, was a big message for the corrupt society. Krishnaprasad Bhattarai, who became the prime minister twice and was the chairman of the Nepali Congress for nearly two decades, had only one box, one water jug and one umbrella.When he left the residence, he had no place to stay. Bhattarai's unmarried relatives were also limited. However, the amount of publicity his death received was unparalleled. Bhattarai's death has taught Nepali society a lesson of simplicity and morality.In this sense, the year 2067 proved to be a year that can convey a message even in grief. Bholanath Jha, an old politician and president of the World Hindu Federation, also died this year. The death of

78-year-old.. He played the role of Minister of Industry, Forest, Education and Health several times.

AUTOBIOGRAPHY OF SHRI BHOLANATH JHA

Bholanath Jha born at Janakpur ,Nepal . He had started his political career from younger age along with his study Acharya 2. Children : Indira Jha (Daughter), Girijanand Jha (Son). Spouce: Manorma Devi Jha

Political Contribution : Defeated his opposition party strong leader Mr.Rudra Giri in parliamentary election on 1973 and after winning election He became very close to His Majesty King Birendra Bir Bikram Shah Dev and he had appointed as a Forest Minister, Health Minister, Education Minister and Udyog as well as Binijya minister in his four years of tenure till 1978. After finish his tenure he was appointed as a member of National Panchayat till 1984 by His Majesty King Birendra Bir Bikram Shah Dev. After finish his tenure he was appointed as member of Jachbujh committee by His Majesty King Birendra Bir Bikram Shah Dev till for another's five years till 1989 .

Religious Contribution: From 2003 to 2012, Bholanath Jha was selected as a President of World Hindu Federation Nepal Chapter by full majority of World Hindu Federation Nepal Chapter committee members. He had started movement to restore Monarchy and Hindu state all over the country because Nepal was declared secular country after dethroned King. 80% of Nepal Population is Hindu but removed Hindu state from constitution and declared Secular state without referendum. From rxpgnews.com NEPAL Nepal Hindu group demands poll on secularism By Sudeshna Sarkar, IANS, Mar 11, 2007 - 2:55:16 PM Kathmandu, March 11 - A controversial Hindu organisation in Nepal, regarded as loyal to King Gyanendra, has started a movement demanding a poll on secularism. The World Hindu Federation -, that was among the first to welcome King Gyanendra's power grab two years ago and urged Hindus worldwide to support the king since he was the monarch of the only Hindu kingdom in the world.

My Respected Guru Dr.Chintamani Nath Yogi Ji

AUTOBIOGRAPHY OF GURU DR.CHINTAMANI NATH YOGI JI

Dr.Chintamani Nath.Yogi leads non-profit organizations in Nepal committed to interfaith compassion, dialogue, and understanding. He established the Hindu VidyaPeeth movement that supports the education of underprivileged and orphaned children while encouraging the values of cultural and religious tolerance. He also established Shanti Sewa Ashram (Peace Service Centre) whose mission is: "To promote peace and spirituality at every level of society, to help the poor and disadvantaged, and to provide a forum for spiritual and peace-oriented organizations in an attempt to unite the effort." In addition, he founded the Youth Society for Peace, an organization that promotes peace, tolerance, and selfless service.

He also directs the Children Study Centre, Children's Peace Home, and the Women's Education/Training Program.

Full Name:Dr. Chintamani Yogi

D.O.B:: 1963 A.D

Birth Place : Dang, Nepal

Current Residence : Kathmandu, Nepal

Education:

• Acharya in Sanskrit, Acharya in Yoga, Masters in English and Phd

ORGANIZATIONAL INVOLVEMENT:

1. Hindu Vidyapeeth-Nepal (Founding principal and Coordinator)
2. Matribhoomi Sevak Sangh (Senior Member)
3. Shanti Sewa Asram (Founding Chairperson)
4. Youth Society for Peace (Patron)
5. Society for Value Education (Founder)
6. Founding Principal- Hindu VidyaPeeth-Nepal.
7. Founding Chairperson- Shanti Sewa Ashram.
8. Patron - Children's Peace Home, Dang (http://www.childrenpeacehome.org.np)
9. Patron- Youth Society for Peace.
10. Patron- Naimisharanya Gurukul, Banepa.
11. Founder - Gorakhanath Sewa Sangh.
12. Chairperson - Global Omkar Federation.
13. Coordinator- Global Network of Religions for Children-Nepal (GNRC) Japan.
14. Coordinator- United Religions Initiative-Nepal (URI) USA.
15. Advisor- Global Communities for Peace, Australia.

Address : Po. Box: 6807, Kathmandu, Nepal; Ph +977- 98510-92805. mail2cmyogi@yahoo.com

AUTHOR'S AUTOBIOGRAPHY

Myself Er.Nirajan Kumar Jha I would like to inform all my well wishers that I wrote the book in January 2017 and the title of the book was "Precast Concrete Technology (Future of Civil Engineering)", this book was highly appreciated by the readers all over the world and after the successful result. I started researching on this to write a second edition of this book to provide more information about precast concrete technology. After my 2 years of research I was able to write the second edition and title of the book is: "Precast, Prestressed and Post Tensioning Technology" Published in July 2023. This book became more popular than the first edition and this book was released across the Globe at one time and many readers sent me mail to write the third edition about eco-friendly building, then I started researching to write the book. And I researched successfully and completed the writing over Book titled " Eco-Friendly Building Construction for Reduction of Global Warming"

Name : Nirajan Kumar Jha

Date of Birth : 31 August 1978

Sex / Marital Status : Male / Married

E-mail: nirajankumar2014@gmail.com ,

Academic Qualifications :

1. Master of Technology in Building Construction and Management from Noida International University.
2. Bachelor Degree of Civil Engineering(First Class , 2005)Visvesvaraya Technological University(VTU) Belgaum, Karnataka, India
3. Diploma in Civil Engineering (First Class, 1999) Bangalora Board, Karnataka, India
4. S.L.C from HMG in Sankat Mochan school (Second Class, 1994)

Key Qualifications:

Summary of Professional Career:

1. Since February 2020 to till date working as a Precast Project Manager in NHRD, India
2. Since 2018 March to 2020 January worked as an Erection Manager in Afriprecast holding group. Rwanda, East Africa.
3. Since February 2018 to March 2018 worked as a Project Engineer in ACICO GROUP, KUWAIT.
4. Since 15th August 2017 to 2018 January worked as a Project Engineer in IPC Group, Dammam, Saudi Arabia.
5. Since 17th June 2015 to 2017 July Worked as a Project Manager in Amrapali Group Precast Division, Uttar Prades, India.
6. Since 19th April 2014 to 30th April 2015 experience as a Erection Engineer in Saudi Arabia based IDEAL PRECAST COMPANY, DAMMAM ,SAUDI ARABIA.
7. Since 05th Nov 2013 to 15th April 2015 worked as a Project Engineer in IPC group , Dammam , Saudi Arabia.
8. Since 19th July 2013 to 30th Oct. 2013 in Saudi based Precast Technologies(Member of Alrajhi Holding Group) as a Project Engineer in National Guard Housing Project(Yanbu), Saudi Arabia.
9. Four years of experience in Qatar based KCIC(KHALID CEMENT INDUSTRIES COMPLEX) FCP Division involved as a Precast Engineer for the

    1. Redco Almana –Horse Breeding Farm Precast Project

    2. Redevelopment of Doha Shiokh Port Project.

    3 Al Ali cricket Stadium Project., Doha, Qatar.

    4 Football Stadium Project, Doha, Qatar.

5. BARWA City Phase 1 Precast Housing Project, Abu Hamour, Doha, Qatar .

10. 1.1years experience in Germany based Bilfinger Berger Qatar W.L.L. as a Precast Engineer in CIP and Pre cast Section for the

construction of BARWA City phase 1 Precast Housing project, Abu Hamour, Doha, Qatar

11. 1.0 year experience in India based Larsen and Toubro Limited ECC Division as a Road Engineer for the construction of SwaroopganjPalanpur Road project, Chitrasani, State- Gujrat, India
12. 2 years experience as a Lecturer in Thapathali engineering college, Tribhuvan University, Kathmandu, Nepal.

Language Skills : Fluency in English, Hindi, Arabic, Maithali and Nepali .

Membership : Nepal Engineering Association Membership No. 4977)

Membership : Nepal Engineering Council (Membership No. 2947)

Membership: Saudi Engineering Council (Membership No. 146893)

Valid Driving license of Doha,Qatar, Saudi Arabia, Nepal, Rwanda & India.

Er. Nirajan Kumar Jha and my wife Mrs. Kamini Kumari Jha

PREFACE

Global warming is now considered to be one of the greatest challenges worldwide. International environmental agreements have been developed in response to climate change since the 1970s. The construction industry is considered one of the main contributors to global warming. In order to mitigate global warming effects, the construction industry has been exploring various approaches to mitigate the impacts of carbon dioxide emissions over the entire life cycle of buildings. The application of different structural systems is considered a means of reducing the carbon dioxide emissions from building construction. The purpose of this research is to assess the environmental performance of three different slab systems during the construction phase. In this study, a process-based life cycle assessment (LCA) method was applied in order to evaluate the level of performance of the three slab systems. The results showed total CO2 emissions of 3,275,712, 3,157,260, and 2,943,695 kg CO2 eq. for the ordinary reinforced concrete slab, flat plate slab, and voided slab systems, respectively. The manufacturing of building materials is by far the main contributor to CO2 emissions, which indicate 3,230,945, 3,117,203, and 2,905,564 kg CO2 eq., respectively. Comparing the building materials in the three slab systems, reinforcing bars and forms were significant building materials to reduce the CO2 emissions in the flat plate slab and voided slab systems. In this study, reinforcing bars were the main contributor to lowering the carbon dioxide emissions in the flat plate slab and voided slab systems. The results of this study show that amongst all the three different slab systems, the voided slab system shows the greatest reduction potential. Moreover, replacing the ordinary reinforced concrete slab system by alternative methods would make it possible to reduce the carbon dioxide emissions in building projects. 3,230,945, 3,117,203, and 2,905,564 kg CO eq., respectively. Comparing the building materials in the three slab systems, reinforcing bars and forms were significant building materials to reduce the CO2 emissions in the flat plate slab and voided slab systems. In this study, reinforcing bars were the main contributor to lowering the carbon dioxide emissions in the flat plate slab and voided slab systems. The results of this study show that amongst all the three different slab systems, the voided slab system shows the greatest reduction potential. Moreover, replacing the ordinary reinforced concrete slab system by alternative methods would make it possible to reduce the carbon dioxide emissions in building projects.

Life cycle assessment (LCA) is a very familiar methodology to measure the environmental effects of any products where all the process associated with the products from cradle to grave was analyzed and the possible emission to environment can be identified. In this study we applied LCA on three traditionally constructed reinforced concrete buildings (one five storied residential building, one three storied office building and one three storied educational building) where no environmental issues were considered during design and construction period. The aim of this research is to evaluate and compare energy consumption and carbon emissions of three different types of buildings from their materialization stage to the end-of-life stage. This paper also describes the step-by-step process of quantifying the overall carbon emission from a building systematically. There is an overview of how emission varies according to buildings material, construction process and objective of buildings. The result shows that the operational phase is mainly responsible for maximum carbon emission due to maximum energy consumption among three phases of life cycle assessment. However, it is also found from the study that the materialization and operation stages together contribute more than 97% of total emissions. Since a huge amount of operational energy is required for commercial building compare to other two buildings, it consumes energy comparatively higher than residential and educational building which results in 13.6% more emission than residential building and 19% more than educational building.

Global warming is considered one of the greatest challenges worldwide. International environmental agreements have been developed in response to climate change since the 1970s. The construction industry is considered to be one of the main contributors to global warming. According to the Inter governmental Panel on Climate Change not only is the construction industry responsible for about 40% of global energy consumption, it is also responsible for 30% of global green house gas emissions per annum. Furthermore, it is common for most countries to construct high-rise buildings and larger-sized facilities, due to the rapid growth of population Considering the materials consumption costs, durability, and the environmental impacts, a more in-depth study is necessary to evaluate the influence of the concrete cover thicknesses over the structure's life cycle. Studies that evaluate the environmental impacts of reinforced concrete slabs are relatively scarce (Paik and Na, 2019). Besides that, there are very few studies investigating the concrete cover thickness as a principal subject, and none thus far evaluate its effect on the structure's life cycle. The cement production and steel are the main contributors to many environmental impacts in structures of conventional concrete (Paik and Na, 2019) and the production of steel and concrete has a higher embodied energy than the other materials as found by Life Cycle Costs and impacts of massive slabs with varying concrete cover systems. In this study, reinforcing bars were the main contributor to lowering the carbon dioxide emissions in the flat plate slab and voided slab systems. The results of this study show that amongst all the three different slab systems, the voided slab system shows the greatest reduction potential. Moreover, replacing the ordinary reinforced concrete slab system by alternative methods would make it possible to reduce the carbon dioxide emissions in building projects.

The devastating environmental changes like global warming, acidification, smog formation, ozone layer and natural resources depletion, biological diversity losses, waste accumulation, etc. are highlighting construction industry as a worldwide agitating sector. More than 40% of world's natural resources and energy are consumed by buildings and about 33% of the total carbon dioxide (CO2) is emitted by buildings . In the traditional method. Green building also known as green construction or sustainable building refers to both a structure and the application of processes that are Environmentally responsible and resource-efficient throughout a building's life-cycle: from planning to design, construction, operation, maintenance, renovation, and demolition. This requires close cooperation of the contractor, the architects, the engineers, and the client at all project stages. The Green Building practice expands and complements the classical building design concerns of economy, utility, durability, and comfort. Green building also refers to saving resources to the maximum extent, including energy saving, land saving, water saving, material saving, etc., during the whole life cycle of the building, protecting the environment and reducing pollution, providing people with healthy, comfortable and efficient use of space, and being in harmony with nature Buildings that live in harmony.

Green building technology focuses on low consumption, high efficiency, economy, environmental protection, integration and optimization.

Leadership in Energy and Environmental Design (LEED) is a set of rating Systems for the design, construction, operation, and maintenance of green buildings which was developed by the U.S. Green Building Council. Other certificate systems that confirm the sustainability of buildings are the British BREEAM (Building Research Establishment Environmental Assessment Method) for buildings and large-scale developments or the DGNB System (Deutsche Gesellschaft for Nachhaltiges Bauen e.V.) which benchmarks the sustainability performance of buildings, indoor environments and districts. Currently, the World Green Building Council is conducting research on the effects of green buildings on the health and productivity of their users and is working with the World Bank to promote Green Buildings in Emerging Markets through EDGE (Excellence in Design for Greater Efficiencies) Market Transformation Program and certification. There are also other tools such as Green Star in Australia, Global Sustainability Assessment System (GSAS) used in the Middle East and the Green Building Index (GBI) predominantly used in Malaysia.

Building information modeling(BIM) is a process involving the generation and management of digital representations of physical and functional characteristics of places. Building information models (BIMs) are files (often but not always in proprietary formats and containing proprietary data) which can be extracted, exchanged, or networked to support decision-making regarding a building or other built asset. Current BIM software is used by individuals, businesses, and government agencies who plan, design, construct, operate and maintain diverse physical infrastructures, such as water, refuse, electricity, gas, communication utilities, roads, railways, bridges, ports, and tunnels.

Although new technologies are constantly being developed to complement current practices in creating greener structures, the common objective of green buildings is to reduce the overall impact of the built environment on human health and the natural environment by:

- Efficiently using energy, water, and other resources
- Protecting occupant health and improving employee productivity (see healthy building)

- Reducing waste, pollution, and environmental degradation Natural building is a similar concept, usually on a smaller scale and focusing on the use of locally available natural materials Other related topics include sustainable design and green architecture. Sustainability may be defined as meeting the needs of present generations without compromising the ability of future generations to meet their needs. Although some green building programs don't address the issue of retrofitting existing homes, others do, especially through public schemes for energy efficient refurbishment, Green construction principles can easily be applied to retrofit work as well as new construction. A 2009 report by the U.S. General services Administration found 12 sustainably-designed buildings that cost less to operate and have excellent energy performance. In addition, occupants were overall more satisfied with the building than those in typical commercial buildings. These are eco-friendly buildings

Table of Contents

CHAPTER 1

Introduction

Global warming is considered one of the greatest challenges worldwide. International environmental agreements have been developed in response to climate change since the 1970s. The construction industry is considered to be one of the main contributors to global warming. According to the Inter governmental Panel on Climate Change not only is the construction industry responsible for about 40% of global energy consumption, it is also responsible for 30% of global green house gas emissions per annum. Furthermore, it is common for most countries to construct high-rise buildings and larger-sized facilities, due to the rapid growth of population Considering the materials consumption costs, durability, and the environmental impacts, a more in-depth study is necessary to evaluate the influence of the concrete cover thicknesses over the structure's life cycle. Studies that evaluate the environmental impacts of reinforced concrete slabs are relatively scarce (Paik and Na, 2019). Besides that, there are very few studies investigating the concrete cover thickness as a principal subject, and none thus far evaluate its effect on the structure's life cycle. The cement production and steel are the main contributors to many environmental impacts in structures of conventional concrete (Paik and Na, 2019) and the production of steel and concrete has a higher embodied energy than the other materials as found by Life Cycle Costs and impacts of massive slabs with varying concrete cover systems. In this study, reinforcing bars were the main contributor to lowering the carbon dioxide emissions in the flat plate slab and voided slab systems. The results of this study show that amongst all the three different slab systems, the voided slab system shows the greatest reduction potential. Moreover, replacing the ordinary reinforced concrete slab system by alternative methods would make it possible to reduce the carbon dioxide emissions in building projects.

The devastating environmental changes like global warming, acidification, smog formation, ozone layer and natural resources depletion, biological diversity losses, waste accumulation, etc. are highlighting construction industry as a worldwide agitating sector. More than 40% of world's natural resources and energy are consumed by buildings and about 33% of the total carbon dioxide (CO2) is emitted by buildings . In the traditional method. Green building also known as green construction or sustainable building refers to both a structure and the application of processes that are Environmentally responsible and resource-efficient throughout a building's life-cycle: from planning to design, construction, operation, maintenance, renovation, and demolition. This requires close cooperation of the contractor, the architects, the engineers, and the client at all project stages. The Green Building practice expands and complements the classical building design concerns of economy, utility, durability, and comfort. Green building also refers to saving resources to the maximum extent, including energy saving, land saving, water saving, material saving, etc., during the whole life cycle of the building, protecting the environment and reducing pollution, providing people with healthy, comfortable and efficient use of space, and being in harmony with nature Buildings that live in harmony. Green building technology focuses on low consumption, high efficiency, economy, environmental protection, integration and optimization.

Global warming is a term used for the observed century-scale rise in the average temperature of the Earth's climate system and its related effects. Scientists are more than 95% certain that nearly all of global warming is caused by increasing concentrations of greenhouse gases (GHGs) and other human-caused emissions. Within the earth's atmosphere, accumulating greenhouse gases like water vapor, carbon dioxide, methane, nitrous oxide, and ozone are the gases within the atmosphere that absorb and emit heat radiation. Increasing or decreasing amounts of greenhouse gases within the atmosphere act to either hold in or release more of the heat from the sun. In order to mitigate global warming effects, the construction industry has been exploring various approaches to mitigate the impacts of carbon dioxide emissions over the entire life cycle of buildings. The application of different structural systems is considered a means of reducing the carbon dioxide emissions from building construction. The purpose of this research is to assess the environmental performance of three different slab systems during the construction phase. In this study, a process-based life cycle assessment (LCA) method was applied in order to evaluate the level of performance of the three slab systems. The results showed total CO2 emissions of 3,275,712, 3,157,260, and 2,943,695 kg CO2 eq. for the ordinary reinforced concrete slab, flat plate slab, and voided slab systems, respectively. The manufacturing of building materials is by far the main contributor to CO2 emissions, which indicate 3,230,945, 3,117,203, and 2,905,564 kg CO2 respectively. Comparing the building materials in the three slab systems, reinforcing bars and forms were significant building materials to reduce the CO2 emissions in the flat plate slab and voided slab systems. In this study, reinforcing bars were the main contributor to lowering the carbon dioxide emissions in the flat plate slab and voided slab systems. The results of this study show that amongst all the three different slab systems, the voided slab system shows the greatest reduction potential. Moreover, replacing the ordinary reinforced concrete slab system by alternative methods would make it possible to reduce the carbon dioxide emissions in building projects.

Comparing the building materials in the three slab systems, reinforcing bars and forms were significant building materials to reduce the CO2 emissions in the flat plate slab and voided slab systems. In this study, reinforcing bars were the main contributor to lowering the carbon dioxide emissions in the flat plate slab and voided slab systems. The results of this study show that amongst all the three different slab systems, the voided slab system shows the greatest reduction potential. Moreover, replacing the ordinary reinforced concrete slab system by alternative methods would make it possible to reduce the carbon dioxide emissions in building projects.

Life cycle assessment (LCA) is a very familiar methodology to measure the environmental effects of any products where all the process associated with the products from cradle to grave was analyzed and the possible emission to environment can be identified. In this study we applied LCA on three traditionally constructed reinforced concrete buildings (one five storied residential building, one three storied office building and one three storied educational building) where no environmental issues were considered during design and construction period. The aim of this research is to evaluate and compare energy consumption and carbon emissions of three different types of buildings from their materialization stage to the end-of-life stage. This paper also describes the step-by-step process of quantifying the overall carbon emission from a building systematically. There is an overview of how emission varies according to buildings material, construction process and objective of buildings. The result shows that the operational phase is mainly responsible for maximum carbon emission due to maximum energy consumption among three phases of life cycle assessment. However, it is also found from the study that the materialization and operation stages together contribute more than 97% of total emissions. Since a huge amount of operational energy is required for commercial building compare to other two buildings, it consumes energy comparatively higher than residential and educational building which results in 13.6% more emission than residential building and 19% more than educational building.

Fig.1

What is Eco-friendly Construction?

An eco-friendly building is a building that uses fewer natural resources and produces fewer emissions and waste. Some green buildings even create a positive impact on the environment around them and contribute to improving the global climate. Eco-friendly buildings are designed to be sustainable throughout their life-cycle.

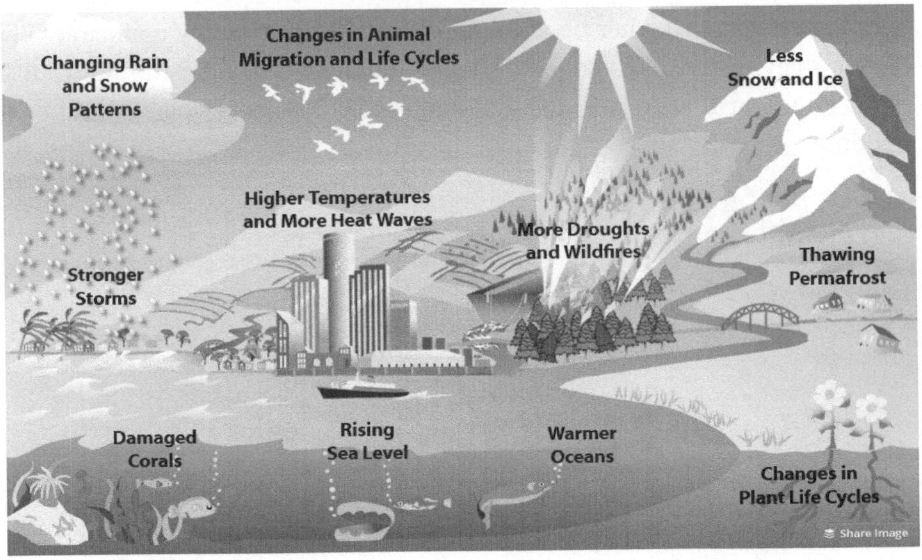

Fig. 2

Green Buildings and Eco-Friendly Construction.

Fig. 3

A. Real Estate

Real Estate and construction accounts for almost 36% of global energy consumption and nearly 40% of greenhouse gas emissions. The demolition and construction generates lot of waste, causes pollution and consumes many natural resources in large numbers. As the world becomes increasingly aware of the need to reduce its environmental footprint, the building industry is being called upon to adopt more eco-friendly practices and use green building techniques and materials.

B. Green building

Green building also known as sustainable building, refers to the practice of designing, constructing, and operating buildings in an environmentally responsible manner. This includes using materials and methods that minimize the environmental impact of the building, as well as ensuring that the building is energy efficient and healthy for its occupants.

C. Use of sustainable materials

One of the most important aspects of green building is the use of sustainable materials. This includes using materials that are made from renewable resources, such as bamboo or cork, or other materials that can be easily recycled, such as reclaimed wood. Additionally, many green building projects use low-toxicity and non-toxic materials, such as natural paint and sealants, to reduce indoor air pollution.

D. Energy Efficiency

Another important aspect of green building is energy efficiency. They are often built with materials that use less energy to create and utilize, such as insulated glass and walls made from concrete. This also includes using energy-efficient appliances, lighting, heating and cooling systems, as well as incorporating solar and other renewable energy sources into the building's design. This not only helps reduce the building's carbon footprint, but it can also save money on energy costs in the long run. According to India's Department of Energy, green buildings can reduce energy consumption by as much as 30-40 percent compared to traditional buildings.

E. Improved indoor Environments

Green buildings focus on creating healthy indoor environments. This includes incorporating natural light, fresh air, and green spaces into the design of the building, as well as using non-toxic materials that do not off-gas harmful chemicals. This helps reduce incidences of asthma, allergies and improve overall health and well-being of the building's occupants.

F. Reduced environmental impact

Another important aspect of green building is the use of sustainable site development practices. This includes preserving natural habitats and ecosystems, protecting water resources, and reducing the heat island effect by using reflective roofing and paving materials.

The planning process involves taking the environment, biodiversity and ecosystems into account by involving all stakeholders and making everything an important part of the green building strategy.

G. Green roofs & walls

One of the most innovative aspects of green building is the use of green roofs and walls. These are essentially green spaces that are integrated into the building's design. Green roofs and walls can help reduce the heat island effect and improve air quality.

H. Lower costs

It might seem that green building is more expensive than traditional building, but it is often not true. In fact, many green building techniques and materials can actually save money in the long run, through energy savings, reduced maintenance costs, improved health and productivity of the building's occupants. They also tend to be much more durable and long-lasting than traditional buildings, due to the use of high-quality sustainable materials and eco-friendly construction methods. This not only helps reduce the overall environmental impact of the building, but it also saves money in the long run by reducing the need for costly repairs, maintenance and renovations.

I. Water conservation and efficiency

Water scarcity is a global problem as it relates to an unsustainable use of water drawn from both surface and underground sources. Increasing demand and depletion of sources for this limited natural resource, makes water conservation and water efficiency an important aspect of the green building strategy. Rain water harvesting gray water recycling, reduced water pressure, installation of plumbing fixtures designed to lower water use, etc are some of the techniques used in green buildings to save water.

J. Reduced waste

Green Buildings draft strategies to reduce the total amount of waste produced, reuse materials that would otherwise be considered waste, and implement waste disposal management through recycling when necessary. Besides, they use fewer resources, leading to less waste and fewer landfill deposits.

Green building is an important step towards reducing the environmental impact of the construction and real estate industry. By using sustainable materials, incorporating energy and water efficiency, healthy indoor environments, and implementing sustainable site development practices, we can create buildings that are better for the environment and better for communities. With increasing awareness and demand for eco-friendly construction, it's an opportunity for builders and developers to innovate and create sustainable, resilient and efficient living spaces.

HOW CAN I MAKE MY BUILDING MORE ECO-FRIENDLY?

There are many things you can do to make an existing building more eco-friendly. Three easy things you can do are:

1. Electricity – Use Less and Supplement with Renewable Energy.
2. Water – Use Less and Recycle Your Grey water.
3. Renovate with Sustainable Materials, Fittings, and Fixtures.

The effects of climate change and global warming.

Climate change destabilises the Earth's temperature equilibrium and has far-reaching effects on human beings and the environment. A distinction is made between direct and indirect effects of climate change. During the course of global warming, the energy balance and thus the temperature of the earth change, due to the increased concentration of greenhouse gases, which has a significant impact on humans and the environment.

It is not scientifically possible to assign individual weather events to the current climate change, however, it can be statistically proven that global warming will increase the probability of extreme weather events. The AR6 synthesis report from 2023 states that "Widespread and rapid changes in the atmosphere, ocean, cryosphere and biosphere have occurred".

Global consequences of climate change.

The consequences of climate change and global warming are manifested worldwide in rising average temperatures, changes in precipitation patterns, rising sea levels and the loss of biodiversity. Increasing extreme weather events and changes to ecosystems and land use harbour increased health risks and have far-reaching economic consequences.

The direct consequences of man-made climate change include:

- rising maximum temperatures
- rising minimum temperatures
- rising sea levels
- higher ocean temperatures
- an increase in heavy precipitation (heavy rain and hail)
- increase in the proportion of violent tropical cyclones
- increase in aridity and drought
- decline in Arctic sea ice and snow cover
- glacier recession and retreat
- thawing permafrost

The indirect consequences of climate change, which directly affect us humans and our environment, include:

- an increase in hunger and water crises, especially in developing countries
- threat to livelihoods from floods and forest fires
- health risks due to increase in frequency and intensity of heat extremes

- economic implications of dealing with secondary damage related to climate change
- increasing spread of pests and pathogens
- loss of biodiversity due to limited adaptability and adaptability speed of flora and fauna
- ocean acidification due to increased HCO_3 concentrations in the water as a consequence of increased CO_2 concentrations
- the need for adaptation in all areas (e.g. agriculture, forestry, energy, infrastructure, tourism, etc.)

Tipping points in the climate system

As the global climate is a highly interconnected system that is influenced by many different factors, the consequences usually result in positive or negative feedback effects. This refers to developments that are self-enhancing due to the occurrence of certain conditions. These tipping points are the crossing of thresholds for which certain consequences can no longer be avoided, even if temperatures were to be lowered again later.

A common example is the ice-albedo feedback, which refers to the melting of the polar caps. According to this, extensive ice surfaces have a cooling effect on the global climate, as a high proportion of radiation is reflected. As a result of the global rise in the average temperature, however, these ice surfaces begin to melt, the ice surfaces shrink and the amount of reflected radiation is reduced. At the same time, the area of land or ocean that has a significantly lower albedo will increase, reflecting less radiation and thus intensifying the actual cause of glacier melt.

Scientists can calculate the tipping points of individual subsystems of the global climate. The higher the global rise in temperature, the more the climate system is affected, so that at a certain point, despite significant efforts, a reversal in the process is no longer possible. Where exactly these tipping points can be found is difficult to predict. Such tipping points are expected for the melting of the polar caps and for the stability of important ocean currents. Other possible tipping points are the disappearance of the Amazon rainforest, the thawing of the permafrost with the release of methane and carbon dioxide, or the acidification of the oceans and the decrease in the absorption capacity for carbon dioxide.

According to the 2021 IPCC report, profound and long-term changes such as melting ice caps, rising temperatures and sea levels or ocean acidification have already been irreversibly set in motion.

I. Climate Impacts on Human Health.

The impacts of climate change include warming temperatures, changes in precipitation, increases in the frequency or intensity of some extreme weather events, and rising sea levels. These impacts threaten our health by affecting the food we eat, the water we drink, the air we breathe, and the weather we experience.

The severity of these health risks will depend on the ability of public health and safety systems to address or prepare for these changing threats, as well as factors such as an individual's behavior, age, gender, and economic status. Impacts will vary based on a where a person lives, how sensitive they are to health threats, how much they are exposed to climate change impacts, and how well they and their community are able to adapt to change.

People in developing countries may be the most vulnerable to health risks globally, but climate change poses significant threats to health even in wealthy nations such as the United States. Certain populations, such as children, pregnant women, older adults, and people with low incomes, face increased risks.

<u>Temperature-Related Impacts:</u>

Warmer average temperatures will lead to hotter days and more frequent and longer heat waves. These changes will lead to an increase in heat-related deaths in the United States—reaching as much as thousands to tens of thousands of additional deaths each year by the end of the century during summer months. These deaths will not be offset by the smaller reduction in cold-related deaths projected in the winter months. However, adaptive responses, such as wider use of air conditioning, are expected to reduce the projected increases in death from extreme heat.

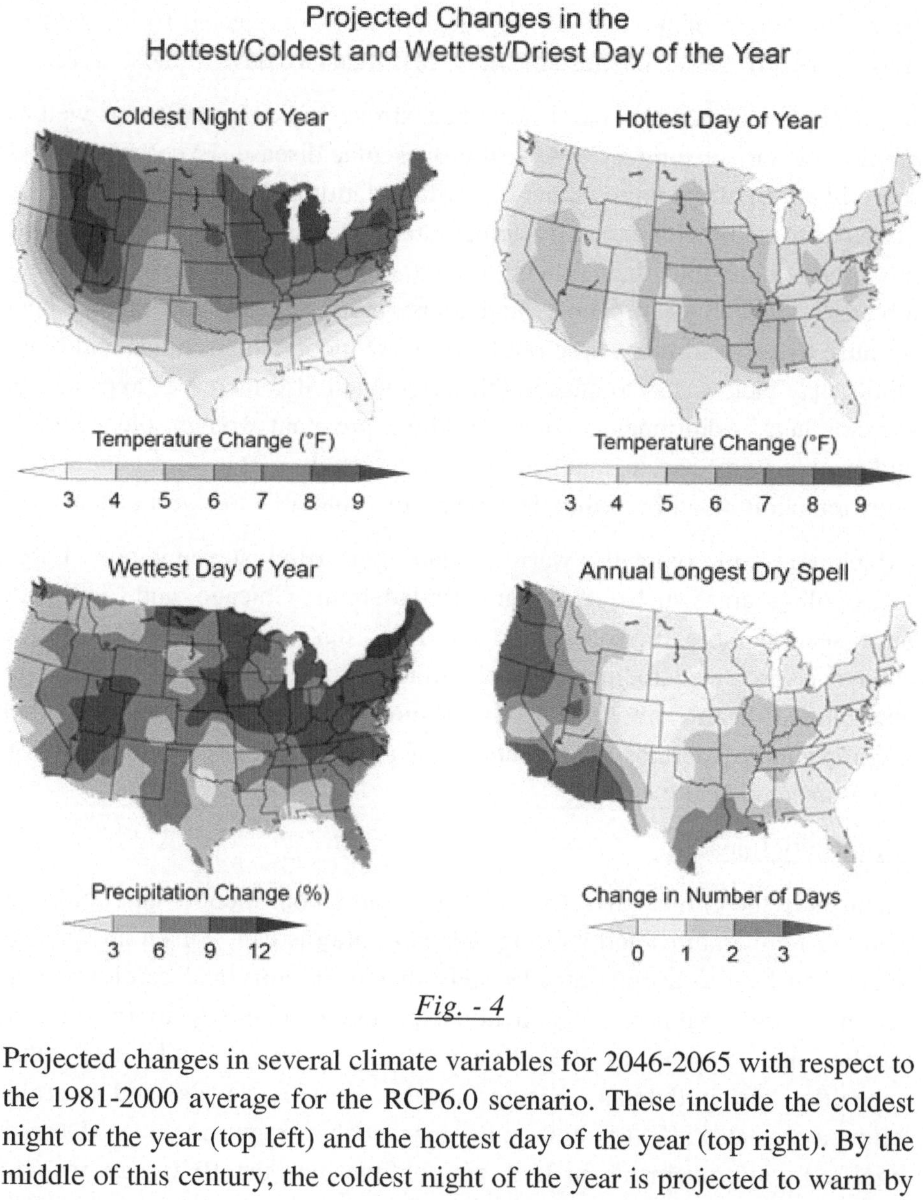

Fig. - 4

Projected changes in several climate variables for 2046-2065 with respect to the 1981-2000 average for the RCP6.0 scenario. These include the coldest night of the year (top left) and the hottest day of the year (top right). By the middle of this century, the coldest night of the year is projected to warm by 6°F to 10°F over most of the country, with slightly smaller changes in the south. The warmest day of the year is projected to be 4°F to 6°F warmer in most areas. Also shown are projections of the wettest day of the year (bottom left) and the annual longest consecutive dry day spell (bottom right). Extreme precipitation is projected to increase, with an average change of 5% to 15% in the precipitation falling on the wettest day of the year. The length of the annual longest dry spell is projected to increase in most areas, but these changes are small: less than two days in most areas.

Exposure to extreme heat can lead to heat stroke and dehydration, as well as cardiovascular, respiratory, and cerebrovascular disease. Excessive heat is more likely to affect populations in northern latitudes where people are less prepared to cope with excessive temperatures. Certain types of populations are more vulnerable than others: for example, outdoor workers, student athletes, and homeless people tend to be more exposed to extreme heat because they spend more time outdoors. Low-income households and older adults may lack access to air conditioning which also increases exposure to extreme heat. Additionally, young children, pregnant women, older adults, and people with certain medical conditions are less able to regulate their body temperature and can therefore be more vulnerable to extreme heat.

Urban areas are typically warmer than their rural surroundings. Large metropolitan areas such as St. Louis, Philadelphia, Chicago, and Cincinnati have seen notable increases in death rates during heat waves. Climate change is projected to increase the vulnerability of urban populations to heat-related health impacts in the future. Heat waves are also often accompanied by periods of stagnant air, leading to increases in air pollution and associated health effects.

Air Quality Impacts:

Changes in the climate affect the air we breathe both indoors and outdoors. Warmer temperatures and shifting weather patterns can worsen air quality, which can lead to asthma attacks and other respiratory and cardiovascular health effects. Wildfires, which are expected to continue to increase in number and severity as the climate changes, create smoke and other unhealthy air pollutants. Rising carbon dioxide levels and warmer temperatures also affect airborne allergens, such as ragweed pollen. Despite significant improvements in U.S. air quality since the 1970s, as of 2014 about 57 million Americans lived in counties that did not meet national air quality standards. Climate change may make it even harder for states to meet these standards in the future, exposing more people to unhealthy air.

Increases in Ozone:

Scientists project that warmer temperatures from climate change will increase the frequency of days with unhealthy levels of ground-level ozone, a harmful air pollutant, and a component in smog.

People exposed to higher levels of ground-level ozone are at greater risk of dying prematurely or being admitted to the hospital for respiratory problems.

Ground-level ozone can damage lung tissue, reduce lung function, and inflame airways. This can aggravate asthma or other lung diseases. Children, older adults, outdoor workers, and those with asthma and other chronic lung diseases are particularly at risk.

Because warm, stagnant air tends to increase the formation of ozone, climate change is likely to increase levels of ground-level ozone in already-polluted areas of the United States and increase the number of days with poor air quality.

Changes in Particulate Matter:

Particulate matter is the term for a category of extremely small particles and liquid droplets suspended in the atmosphere. Fine particles include those smaller than 2.5 micrometers (about one ten-thousandth of an inch). Some particulate matter such as dust, wildfire smoke, and sea spray occur naturally, while some is created by human activities such as the burning of fossil fuels to produce energy. These particles may be emitted directly or may be formed in the atmosphere from chemical reactions of gases such as sulfur dioxide, nitrogen dioxide, and volatile organic compounds.

1. Inhaling fine particles can lead to a broad range of adverse health effects, including lung cancer, chronic obstructive pulmonary disease (COPD), and cardiovascular disease.
2. Climate change is expected to increase the number and severity of wildfires. Particulate matter from wildfire smoke can often be carried very long distances by the wind, affecting people who live far from the source of this air pollutant.
3. Older adults are particularly sensitive to short-term particle exposure, with a higher risk of hospitalization and death. Outdoor workers like firefighters can also have high exposure.

4. Due to the complex factors that influence atmospheric levels of fine particulate matter, scientists do not yet know whether climate change will increase or decrease particulate matter concentrations across the United States. Particulate matter can be removed from the air by rainfall, and precipitation is expected to increase in quantity though not necessarily frequency. Climate-related changes in stagnant air episodes, wind patterns, emissions from vegetation and the chemistry of atmospheric pollutants will also affect particulate matter levels.
5. The higher concentrations of ozone due to climate change may result in tens to thousands of additional ozone-related illnesses and premature deaths per year by 2030 in the United States, assuming no change in projected air quality policies.

Changes in Allergens and Asthma Triggers:

Allergic illnesses, including hay fever, affect about one-third of the U.S. population, and more than 34 million Americans have been diagnosed with asthma. Climate change may affect allergies and respiratory health.The spring pollen season is already occurring earlier in the United States for certain types of plants, and the length of the season has increased for some plants with highly allergenic pollen such as ragweed. In addition to lengthening the ragweed pollen season, rising carbon dioxide concentrations and temperatures may also lead to earlier flowering, more flowers, and increased pollen levels in ragweed.

Vectorborne Diseases:

Vectorborne diseases are illnesses that are transmitted by disease *vectors*, which include mosquitoes, ticks, and fleas. These vectors can carry infectious pathogens, such as viruses, bacteria, and protozoa, from animals to humans. Changes in temperature, precipitation, and extreme events increases the geographic range of diseases spread by vectors and can lead to illnesses occurring earlier in the year.

1. The geographic range of ticks that carry Lyme disease is limited by temperature. As air temperatures rise, ticks are likely to become active earlier in the season, and their range is likely to continue to expand northward. Typical symptoms of Lyme disease include fever, headache, fatigue, and a characteristic skin rash.

2. Mosquitoes thrive in certain climate conditions and can spread diseases like West Nile virus. Extreme temperatures—too cold, hot, wet, or dry—influence the location and number of mosquitoes that transmit West Nile virus. More than three million people were estimated to be infected with West Nile virus in the United States from 1999 to 2010.

The spread of climate-sensitive diseases will depend on both climate and non-climate factors such as land use, socioeconomic and cultural conditions, pest control, access to health care, and human responses to disease risk. The United States has public health infrastructure and programs to monitor, manage, and prevent the spread of many diseases. The risks for climate-sensitive diseases can be much higher in poorer countries that have less capacity to prevent and treat illness.

West Nile virus is maintained in transmission cycles between birds (the natural hosts of the virus) and mosquitoes. Human infections can occur from a bite of a mosquito that has previously bitten an infected bird. Warmer winters, longer frost-free season, and earlier spring arrival may influence the migration patterns and fledgling survival of birds that are the natural host of West Nile virus. In addition, rising temperature, changing precipitation patterns, and a higher frequency of extreme weather events are likely to influence the distribution and abundance of mosquitoes that transmit West Nile virus. Source: USGCRP (2016)

Water-Related Illnesses:

People can become ill if exposed to contaminated drinking or recreational water. Climate change increases the risk of illness through increasing temperature, more frequent heavy rains and runoff, and the effects of storms. Health impacts may include gastrointestinal illness like diarrhea, effects on the body's nervous and respiratory systems, or liver and kidney damage.

Climate impacts can affect exposure to waterborne pathogens (bacteria, viruses, and parasites such as *Cryptosporidium* and *Giardia*); toxins produced by harmful algal and cyanobacterial blooms in the water; and chemicals that end up in water from human activities.

Changing water temperatures mean that waterborne *Vibrio* bacteria and harmful algal toxins will be present in the water or in seafood at different times of the year, or in places where they were not previously threats.

Runoff and flooding resulting from increases in extreme precipitation, hurricane rainfall, and storm surge will increasingly contaminate water bodies used for recreation (such as lakes and beaches), shellfish harvesting waters, and sources of drinking water.

Extreme weather events and storm surges can damage or exceed the capacity of water infrastructure (such as drinking water or wastewater treatment plants), increasing the risk that people will be exposed to contaminants.

Water resource, public health, and environmental agencies in the United States provide many public health safeguards to reduce risk of exposure and illness even if water becomes contaminated. These include water quality monitoring, drinking water treatment standards and practices, beach closures, and issuing advisories for boiling drinking water and harvesting shellfish.

Food Safety and Nutrition:

Climate change and the direct impacts of higher concentrations of carbon dioxide in the atmosphere are expected to affect food safety and nutrition Extreme weather events can also disrupt or slow the distribution of food.

1. Higher air temperatures can increase cases of *Salmonella* and other bacteria-related food poisoning because bacteria grow more rapidly in warm environments. These diseases can cause gastrointestinal distress and, in severe cases, death. Practices to safeguard food can help avoid these illnesses even as the climate changes.
2. Climate change will have a variety of impacts that may increase the risk of exposure to chemical contaminants in food. For example, higher sea surface temperatures will lead to higher mercury concentrations in seafood, and increases in extreme weather events will introduce contaminants into the food chain through storm water runoff.
3. Higher concentrations of carbon dioxide in the air can act as a "fertilizer" for some plants, but lowers the levels of protein and

essential minerals in crops such as wheat, rice, and potatoes, making these foods less nutritious.

4. Extreme events, such as flooding and drought, create challenges for food distribution if roads and waterways are damaged or made inaccessible.

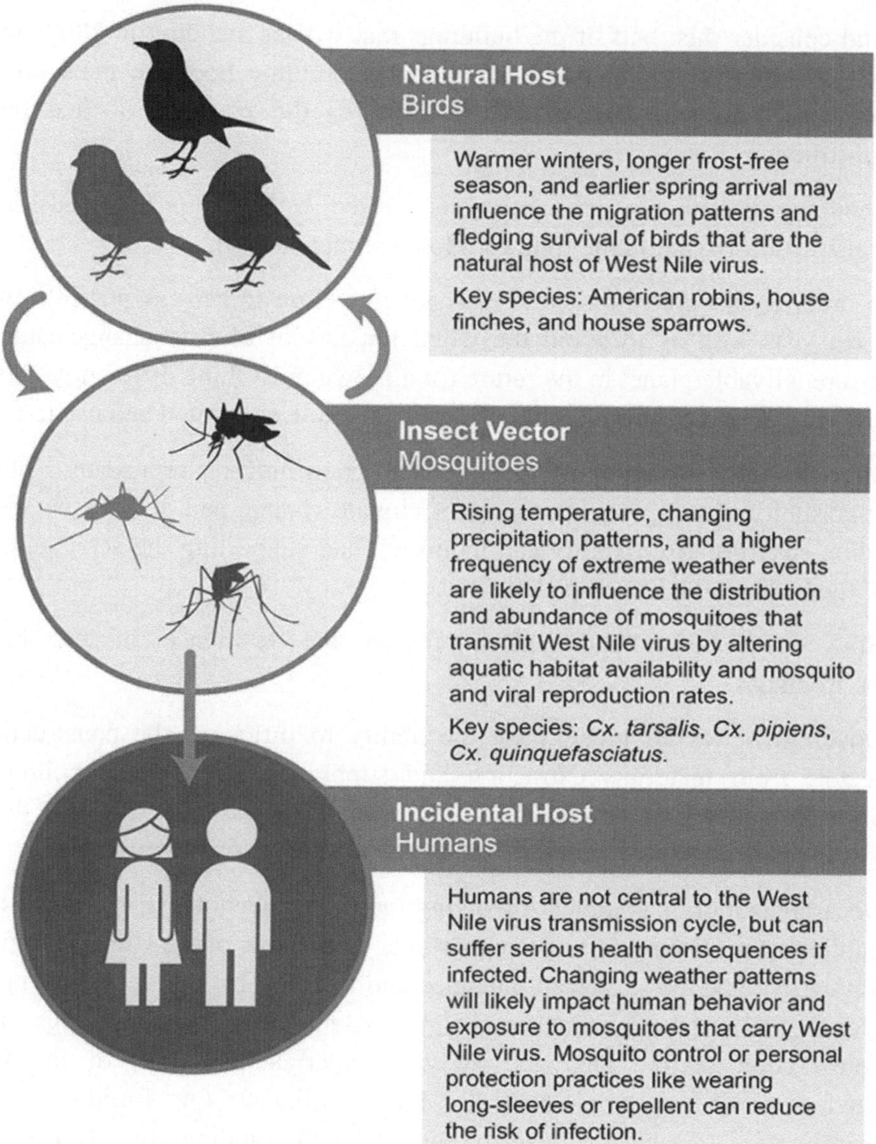

Fig. - 5

Decarbonizing Buildings is vital to climate action.

Infrastructure is responsible for 79 per cent of all greenhouse gas emissions – and 88 per cent of all adaptation costs. In Europe alone, for example, the buildings sector represents almost 40 per cent of Europe's energy demand – 80 per cent of it from fossil fuels.

And consider this: half of the buildings that will be standing in 2060 have not yet been built. Thanks to an expected building boom in the coming decades, Asia and Africa will account for the majority of that new construction.

Reducing greenhouse gas emissions caused by building materials and construction is vital in the fight for global climate action.

In order to cut greenhouse gas emissions as close to zero as possible (net zero) – as well as to avoid the worst impacts of climate change and to ensure a livable planet in the future for all – we must think differently about how our cities and communities are planned, built, renovated and used.

Improving the sustainability of the construction industry represents a huge opportunity to mitigate the impact of climate change and achieve net zero while ensuring affordability and inclusion, and supporting the achievement of the Sustainable Development Goals.

How by decarbonizing and improving the resilience of the built environment.

Governments at all levels have the ability to influence the construction sector: from materials extraction and supply chains to their buildings portfolio, public infrastructure, land use planning – and more. In short, they, together with other stakeholders are key to net zero infrastructure.

Decarbonization is complex, involving multiple stakeholders from both the public and private sectors. To support a key objective of the Climate Action Pathways, UNOPS produced guidance and a toolkit for the development of global and regional climate action roadmaps for the buildings and construction sector. This activity was undertaken as part of the UN Environment Programme-hosted Global Alliance for Buildings and Construction (GlobalABC) and completed with funding from Germany's Federal Ministry of Economic Cooperation and Development.

The global and regional roadmaps provide comprehensive frameworks for national, regional and local governments to achieve zero-emission, efficient and resilient buildings, helping to decarbonize the buildings and construction industry by 2050.

The roadmaps support governments with setting and including ambitious targets related to buildings and construction in their Nationally Determined Contributions (efforts by each country to reduce national emissions and adapt to the impacts of climate change) and country strategies.

The roadmap development guidance and supporting tools focus on five action areas (strategic priorities, urban and spatial development, new buildings, existing buildings and construction supply chain) and four objectives (zero embodied carbon, zero operational carbon, resilience, and inclusion and wellbeing). It includes several suggestions for meeting decarbonization goals, such as promoting the use of low-carbon materials, increasing access to clean energy, and building resilience for buildings and communities, among other recommendations.

Environment Friendly Construction Methods.

As society turns its attention toward our evolving climate crises and its effects on a global scale, architects and construction managers have evolved their practices with it. To reduce waste and protect the environment, there is now a higher value on sustainable and green building methods. There are even highly desired certifications like LEED that honor the design, construction, operations, and maintenance of high-performing, resource efficient, healthy, and cost-effective buildings. With the new advances in technologies, materials, and practices, designing and constructing has never been more focused on sustainable methods. Here are 15 ways that architects and construction managers have turned the tide on environmentally friendly and sustainable construction.

A. Prefabricating Materials in Controlled Environments.

Advantages of Hiring Building Information Modeling Services Constructing as much of a structure in a controlled environment as possible has been proven to improve the quality of buildings and results in less trash. Mechanical contractors have opted to use Building Information Management (BIM) systems that accurately cut sheet metal for ductwork in controlled environments rather than constructing the materials needed on site. Due to this innovation, external factors like cold or hot weather can no longer create shape-changing problems on the sheet or other damage. With precise cutting systems, it also allows contractors to use wood framing for buildings as high as five stories because the cuts are more accurate and in turn, can support more weight.

B. Construction Waste Management.

Waste: Construction Waste in a construction site, there are multiple different types of materials being used at a time. Much of these materials get dumped in landfills and never used again. Rather than throwing away usable materials like concrete and structural steel into landfills, they can be recycled for other construction jobs. Carpet, flooring, and ceiling materials can also be reused and in turn, create a substantial saving opportunity for other sites. Companies that reuse materials and sustainable materials can also avoid the additional costs of disposing toxic materials from the site. The process of throwing away waste has also significantly changed. Job sites now use one trash bin for all types of waste, and use haulers with pickers to separate materials and determine what can be reused or not.

C. Passive Solar Design.

Passive Solar Heating for Survival at Self-Reliance Association a construction method that optimizes solar energy use for its building is following the passive solar concept. Passive solar design works with the sun's heat energy to heat the building while also blocking the sun in the hotter summer months and keeping the building cool. Due to the sun rising in the east and set in the west, if a structure has windows on the southern facing side of any building, it can potentially reach the sun's heat energy throughout the day. The passive solar design utilizes energy from the sun, using its predictable movements through the seasons to heat and cool a building without any additional mechanical devices.

D. Environmentally-Friendly Insulation Materials.

Eco-Friendly Insulation Alternatives For A Green Home and Miller, Insulation materials are used to retain heat during the winter by restricting airflow, and also reduces the cost of heating and cooling throughout the year. A majority of buildings use insulation materials that can contain toxic chemicals such as petrochemicals and toxic adhesives. There are now sustainable insulation materials that present a range of safe, natural, and readily available materials. Construction companies argue that these materials are more costly, but some are more affordable than their commercial counterparts. A few examples of green insulation materials include sheep wool, recycled cotton from shreds of discarded clothing, Icynene, and cellulose, made from recycled newsprint and other paper.

E. Active Solar Power.

Tesla roof tiles turn solar shingles into a complete solar roof ©Elemental Green

In addition to passive solar energy, active solar power is the use of functional solar systems that absorb the sun's radiation to cater for heating and electricity provision. It reduces the need for the use of electricity or gas. There are many advances in solar technologies than the generic rectangular square we've grown accustomed to. Solar roof shingles are small, custom shingles designed to blend in almost unnoticeably with traditional roof tiles. Not only are these shingles aesthetically appealing, but they function both as solar panels and as durable roof tiles, protecting roofs from the natural elements as it absorbs sunlight for energy use.

F. Electric Smart Glass.

Halio is a smart glass that can darken on demand within 3 minutes at Hardware zone. Electronic Smart Glass is one of the latest technologies in sustainable construction. The electronic smart glass shuts out the harsh heat of solar radiation, reflecting approximately 98% of sun rays. The smart glass uses tiny electric signals to charge the windows, changing the amount of solar radiation it reflects. It is incorporated into the building's control system, allowing users to choose the amount of solar radiation to block. The glass simply changes from light to dark (clear to opaque) at the push of a button, optimizing user compatibility. With this technology, any type of building can save a lot on heating, ventilating, and air conditioning costs.

G. Using Locally Sourced Materials.

Rickenbacker welcomes weekly Asiana Cargo flights at AIR CARGO WEEK

Depending on the project, materials are sometimes transported over long distances, spending a lot of energy and money in the transportation phase to get the material to the site. With green construction, there is an emphasis on the use of locally-sourced materials. This is meant to reduce the carbon footprint of transportation and creates a far more efficient construction process. Moreover, using materials that are available in any given city or town, construction companies can help support and grow local businesses by buying materials from them. Aesthetically speaking, using locally found materials also better integrate into the surrounding environment.

H. Water Efficient Gray water Recycling System.

Allegheny County Eyes Code Update for Home Gray Water Systems.The Allegheny Front Construction companies can also use water-efficient technologies as part of their sustainable construction projects. One of these is the Gray water recycling system, which is the process of reusing water that has already been used in instances like laundry or bathwater and has it serve another purpose such as garden irrigation, toilet flushing, and car washing. These new technologies can re-use and enhance efficient water supply systems. The Gray water system could lower water and sewage bills, reduce overall water consumption, and help deal with combined sewer overflow issues.

I. Water Efficient Dual Plumbing System.

Out When Facing Plumbing Problems In Kutztown Pennsylvania, Contact Robinson Plumbing Pipes Wearing.There are other water-efficient technologies that construction teams can use as part of their sustainable construction. One of these includes dual plumbing, a system designed to run potable water and reclaimed water through the building in separate piping systems. The two types of water never meet. Potable water, water used for drinking and washing is properly sanitized and filtered for the safety of the residents. Reclaimed water, such as toilet and irrigation water, is also filtered to avoid developing any harmful bacteria. Through dual plumbing, no water is wasted in the sewer system.

J. Using Biodegradable Materials.

A guide to eco and natural paints the use of biodegradable materials is an environmentally friendly means of developing sustainable construction projects. A majority of traditional construction methods lead to the accumulation of waste products and toxic chemicals, which can take hundreds of years to degrade. Biodegradable materials such as organic paints, therefore, can help reduce the negative impacts on the environment as they easily breakdown without releasing toxins or other harmful substances. There are other biodegradable materials available for building foundation, walls, and insulators that are also part of sustainable construction technologies and can reduce a project's harm to the environment.

K. Rammed Earth Material.

Rammed earth brick is an ancient building material that can be dated back to early man-made constructions and can now be found in luxury homes as dramatic and durable walls that resemble sedimentary rock. This low-carbon technique uses soil and binder, placing them together in layers and then applying pressure to create a hard and durable surface. Because it uses sustainably sourced raw materials, it is a green alternative to conventional concrete masonry. Rammed earth walls or floors can be used as thermal storage, absorbing sunlight during the day and releasing the warmth in the colder evenings.

L. Increased Green and Multiuse Space.

With space in cities increasingly limited, developers and architects must plan to optimize green space and create more multipurpose areas. Instead of building a commercial area solely with one purpose, it would be more proactive to apply a multipurpose, and a mixed-use approach to that space. Green spaces on rooftops or in other parts of a building can also add on to this mixed-use approach. This not only provides a multi use space for users but also helps to improve the air quality in any given building. Now, even with the limited space, the structure can serve multiple uses instead of having to be built individually, reducing materials and environmental disruption.

M. Proper Site Selection

Proper site selection and prevention of pollution on the construction site can be a decisive factor in determining how environmentally conscious a project will be. Developers and designers should avoid constructing on prime farmland, in a flood plain, on threatened animal habitat, and on areas close to wetlands. Doing so can help the environment by being as minimally disruptive as possible during the construction process, avoiding pollution and waste to inhabit those areas, and eventually when the project is complete.

N. Harnessing Wind Power

Self-powered buildings are one of the main goals in sustainable design. These buildings are built in a way they are able to generate sufficient power to support their own energy needs and redirect surplus power back into the power grid, Wind power is clean and renewable energy that can be used in skyscrapers where wind turbines are attached to the rooftops, using the heavy air currents at that higher altitude to generate power. Wind turbines are easy to manage, needing very little maintenance after it is installed, and it can help maintain indoor air quality and reduce utility investments.

O. Durable Wood Materials

The wood material is appearing in construction in never before seen ways. Wood bark comes from trees that have already been cut. Previously, tree bark would be discarded as debris or burned. But now, it can be used as a highly durable, economical, and sustainable option for siding and shingling. Another example of highly durable wood material is wood that has been treated with acetylation. Acetylation creates a product resistant to rot, improved in thermal insulation with better dimensional stability, and a longer lifespan. It results in a beautiful and certified non-toxic material that can reduce costs in insulation and material costs as well as safety concerns.

What is Sustainable Construction?

Sustainable construction is the practice of creating a healthy environment that's based on ecological principles. According to Professor Charles J. Kibert, sustainable construction focuses on six principles: "conserve, reuse, recycle/renew, protect nature, create nontoxic and high quality. The goal is to reduce the industry's impact on the environment by utilizing sustainable development practices, employing energy efficiency and taking advantage of green technology. Although most business sectors are making strides to become more sustainable, the construction sector is unique because it can significantly affect how sustainable practices are applied. This is because of the large amounts of materials and energy that the industry uses.

What are the sustainable Building materials?

Sustainable Building Materials

One of the best ways to practice sustainability in construction is through the materials that are used. A new generation of stronger, lighter and revolutionary building materials can help solve many problems in the industry as well as push current practices to be more sustainable. These materials have the added benefit of protecting the environment by reducing the carbon footprint of the buildings that use these materials. They promote a cleaner Earth and a future of sustainability while also being aesthetically appealing and much more efficient.

A. Bamboo

Fig. 6

Thanks to its renewable and versatile characteristics, bamboo is one of the most sustainable construction materials. Although bamboo is a type of grass, it has strength similar to wood while still being flexible. Construction projects can utilize bamboo in various applications like supporting concrete, scaffolding, roofing and building other structures.

Benefits:

- Fast-growing renewable material

- Cost-effective and eco-friendly to grow and harvest
- Absorbs more CO2 than trees, so a growing demand for bamboo could lead to cleaner air in surrounding environments.

B. Recycled Plastic

In 2021, the total world population created 139 million tons of single-use plastic with the majority of it ending up in landfills or polluting waterways. With new advancements in construction and technology, we can now recycle plastic to create building materials, such as plastic sheets, bricks and lumber.

Benefits:

1. Cuts down on waste that ends up in landfills, waterways and other polluted areas.
2. Recycled plastic does not require the same toxic preservative sealants that lumber needs.

C. Laminated Timber

Also known as mass timber, laminated timber is prefabricated timber that has a higher strength and water resistance than traditional timber. Its strength and water resistance are important factors, as it has the potential to replace steel and concrete. Both steel and concrete produce a much higher carbon footprint than mass timber during their production processes. Laminated timber can support structures such as beams and columns. You might also see it used in roofing and flooring.

Benefits:

- Makes a smaller carbon footprint during production when compared to steel and other structural materials
- Faster to work with to cut down on CO2 emissions associated with construction.

D. Stone

Stone occurs naturally in the earth and can be used both as a building material and as home furnishings, such as titles and countertops. Stone is both durable and low maintenance and, thanks to its versatility, it produces little to no waste when used in construction projects. Since it's a naturally occurring material, it often doesn't require factory production, which cuts down on CO_2 emissions.

Benefits:

- A lifelong investment that won't need replacing
- Is recyclable for use on other projects or in the making of roadbeds for little to no waste

E. Cob

Cob is a mud mixture made from natural materials such as soil, straw, sand and lime. Cob most frequently appears in the construction of residential buildings or as a replacement for concrete structures. The making of cob is inexpensive and produces less CO_2 than producing concrete. However, the material has a few disadvantages as it takes longer to build with and is more susceptible to mold if humidity levels are too high.

Benefits:

- All-natural and economically-friendly replacement to concrete in smaller structures
- Prevents heat transfer to keep heating and energy costs low and shrink the home's carbon footprint

F. Cork

Cork is currently common in European construction and is slowly making its way to the United States to insulate homes and other buildings. Cork comes from the cork oak tree. The material is harvested by hand from the tree's bark and, most importantly, doesn't require killing the tree. As a renewable and recyclable resource, cork is an eco-friendly replacement for traditional insulation that requires manufacturing.

Benefits:

- Cork is a renewable resource that doesn't harm the environment to harvest
- Production creates a smaller carbon footprint than traditional human-made insulating materials

- Has mold-resistant properties

G. Adobe Brick

The use of adobe brick dates back centuries and is popular in the Middle East and in the Americas. Adobe is a mixture of clay and straw that people use to make bricks for the construction of homes and other structures. The main appeal of adobe is that it uses naturally occurring materials from the earth and takes less energy to produce.

Benefits:

- Contains natural insulating properties that keep indoor temperatures consistent for lower energy bills and a smaller carbon footprint
- Adobe is recyclable time and time again, leaving no waste behind.
- Consists of sustainable materials from the earth

H. Reclaimed Wood

Due to its ease of use and natural beauty, wood is the most popular building material out there. Unfortunately, it often goes to waste during a deconstruction project. Reclaimed wood technology can now deconstruct lumber from older structures while preserving its integrity. Contractors and carpenters can use reclaimed wood in new building projects or in home furnishing and decor.

Benefits:

- Reduces the need to cut down and harvest trees for new wood
- An effective step in reducing deforestation

I. Precast Concrete

While concrete tends to be a factor in CO2 emissions created at construction sites, it is still a much-needed material. That's where precast concrete comes to play. Precast concrete is factory-made in exact measurements and then shipped to the construction site. While you may not notice the difference between regular and precast concrete, you may see precast concrete featured in structures like bridges, foundations, parking garages and, in some cases, entire buildings.

Benefits:

- Exact-batch technology during the production process leads to less waste
- Can adapt to various climates without cracking or sustaining other damages

J. Mycelium

Mycelium is the thin fibers from fungi that run underneath the ground as if they are roots. When harvested and dried, mycelium becomes an extremely durable, water, mold and fire-resistant building material. Dried mycelium can combine with other materials, like sawdust and demolition waste, to create bricks for building structures. While not widely used today, mycelium is an organic and renewable resource capable of revolutionizing the construction industry as we know it.

Benefits:

- Renewable and earth-friendly resource
- Completely compostable and leaves behind no waste

K. Sheeps Wool

Sheep's wool is a renewable and natural resource that can be harvested and used without hurting the animal. While we typically use wool in the making of clothing and other textiles, it's also useful to insulate buildings. Unlike human-made insulation, sheep's wool is natural, non-toxic and mold resistant.

Benefits:

- Lasts longer over time, leading to less waste
- More effective than its man-made counterparts
- Provides more energy savings to those who use it

L. Pollution-Absorbing Brick

With growing environmental concerns, air pollution is one of the main problems that has proved difficult to solve. While other sustainable building materials are effective ways to

cut down on C02 emissions, pollution-absorbing brick seeks to neutralize emissions. These double-layered bricks have porous blocks that allow air to pass through as it filters out pollutant particles — course and fine. While pollution-absorbing bricks are rarely seen in construction today, this futuristic material can be a sustainable, air-filtering alternative to the use of earth brick.

Benefits:

- Self-sustaining ventilation system
- Innovative solution for greener construction materials and practices.
- Growing cities can use this technology to provide cleaner air to its residents

M. <u>3D-Printed Concrete</u>

We know that concrete is a building material that's not going anywhere any time soon. While there are a few viable alternatives to concrete, there are also ways to make the production of concrete more sustainable. One of which is the use of 3D printing. 3D-printed concrete allows the contractors to digitally design any shape and use concrete to "print" it.

Benefits:

- Time, energy and money-saving
- The ability to print on demand leaves no waste behind
- Save energy and money on transporting materials

N. <u>Cordwood</u>

Cordwood is a building method that uses stacked short logs, similar in shape to firewood, to build a wall. Cob or mortar is used between the wooden logs to secure everything together. The cordwood technique is typically seen in homes and provides a rustic cottage look to the overall structure. Cordwood offers natural insulation and can use local materials to save energy and money on transportation.

Benefits:

- Inexpensive and easy to construct
- Saves time and energy on home construction

- The mixture of cob and wood prevents heat transfer for lower energy consumption

O. Recycled Tires

Until recently, rubber tires weren't recycled and were left cluttering landfills. Thankfully, experts have found that the rubber found in tires provides durability, flexibility and insulation, all of which are important factors in building materials. Some sustainability-focused homes, known as Earthship homes, use recycled tires filled with sandbags as insulation. Others are working on ways to make rubber masonry blocks with recycled tires that builders can use for new structures.

Benefits:

- A sustainable way to keep tires out of a landfill
- A recycled material that can convert into home insulation

P. Newspaper Wood

Although paper is one of the easiest materials to recycle, paper waste is a growing concern. With new innovations and advances in technology, companies like Newspaper Wood seek to turn paper back into wood. While newspaper wood is not yet usable in construction projects, there are promising results in furniture and home decor.

Benefits:

- Keeps paper waste out of landfills
- Prevents the overconsumption of new lumber
- Slows deforestation

Q. Plant-Based Polyurethane Rigid Foam

Rigid foam has been around for a few decades as an effective insulator for homes and commercial buildings. However, a key compound in the original version of rigid foam was found to be harmful to the environment. That's where rigid foam's newer, eco-friendly counterpart is helpful. Plant-based polyurethane rigid foam uses a mixture of bamboo, hemp and kelp that is great wall insulation. This improved rigid is also good for furniture and even surfboards.

Benefits:

- Protects against mold and pests
- Has a longer life span than other traditional insulation methods
- A high-quality thermal insulation that saves energy and money on cooling and heating

R. Straw Bales

The use of straw in construction dates back centuries before our current day, but it still proves to be one of the most sustainable building materials readily available today. Like many of the other materials already mentioned in this article, straw is another renewable resource that works well as insulation. As long as it is properly protected from moisture, straw bale insulation can last for years on end.

Benefits:

- Improves air quality and absorbs carbon dioxide
- Produces low emissions, especially when compared to human-made insulation
- A fast-growing and all-natural reusable insulation material
- Biodegradable and leaves no waste behind

S. Recycled Glass

Glass is one of the hardest materials to recycle, leaving it discarded in landfills and high-pollution areas. As a step in the right direction, recent studies are finding that construction projects can use waste glass to imitate natural aggregates, like sand, gravel and crushed stone. Cement makers can also use recycled glass in their mixture. While using recycled glass is not a common practice in construction, this is an exciting new development that could benefit the environment and our landfills.

Benefits:

- Keeps discarded glass from cluttering landfills, waterways and polluted areas
- Provides a replacement for natural resources in construction

T. Recycled Steel

As previously mentioned, producing steel can be harmful to the environment, yet it continues to be necessary for building frameworks and supporting entire structures. One of the best qualities of steel is that it has an unlimited life cycle, which means it can be recycled repeatedly and never lose its quality or durability. Seeking out recycled steel for new construction projects saves energy while effectively reducing a project's carbon footprint.

Benefits:

- Slows the production of new steel, which emits large amounts of CO_2
- Steel is the easiest material to recycle without losing its quality or value.

Explain Sustainable Construction Methods.

Sustainable Construction Methods

Sustainable construction goes beyond the materials contractors use in their projects. Certain construction practices and methods are proven to be more earth-friendly and enhance sustainable efforts. Sustainable construction methods include, but are not limited to:

1. Using exact measurements to cut and produce materials and minimize waste
2. Improving waste management by recycling materials
3. Constructing green buildings
4. Refurbishing old builds rather than building new ones
5. Managing the CO_2 transmissions created by construction sites
6. Encouraging workers to use eco-friendly practices onsite Examples of these practices include recycling food packaging or using reusable containers, not smoking, minimizing paper usage, etc.
7. Conserving energy

8. Using the sustainable materials above

What Are the Benefits of Sustainable Construction?

Sustainable building isn't just good for the environment, although that is a fantastic reason to adopt sustainable practices. There are many benefits to adopting eco-friendly methods in the construction industry, such as:

1. <u>Promotes Healthier Living</u>

Construction projects that develop green buildings aren't only beneficial to the environment; they also provide many psychological benefits to the people inside them. For example, in an office building, cognitive function scores rose by 61 percent. It was also reported that employees were 44 percent better at making decisions that achieve workplace goals. In green hospital buildings, 56 percent were satisfied with the cheerfulness of the hospital after the green renovations. Plus, Seasonal Affective Disorder was reduced.

2. <u>Reduces Waste</u>

The reduction of construction waste is also a beneficial side effect of building more green buildings. By their nature, they already use fewer resources, relying on recycled and renewable materials along with more sustainable construction methods. The use of sustainable materials is also beneficial to overall human health as paint, industrial cleaning products and building materials can be dangerous for human health.

3. <u>Boosts the Economy</u>

Sustainable construction can also provide many jobs and boost the economy. As climate change devastates the world, efforts to combat its effects have increased, resulting in an increased demand for construction workers and a hike in construction jobs.

4. <u>Promotes Sustainability</u>

Sustainable construction also promotes sustainability and efficient energy use. With renewable energy construction on the rise, coupled with sustainable construction methods, more people are beginning to see the importance and efficiency of using sustainable methods. It also sends a clear message to the industry and everywhere else: sustainability is viable and important.

What Are the Challenges of Sustainable Construction?

While sustainable construction comes with a multitude of benefits, switching to a more sustainable process takes time and preparation to utilize the best practices. Training needs to be implemented in order to start practicing sustainable methods which takes time and money. Another obstacle that many companies may come across is the actual principal cost of sustainable construction. The general consensus is that sustainable construction comes at a premium, with a higher cost than demand.

Nevertheless, as interest in sustainability continues to rise, construction firms are beginning to make sustainable efforts, which we can see through the rise in green building activity.

Importance of Sustainable Construction.

Whether it's the price tag for the materials, the training that goes behind it or resistance to adapting to new methods ('why fix if it ain't broke," as the old saying goes), there is some pushback on green construction. Now that the effects of climate change are already felt across the globe. More owners and developers, both public and private, are turning to a greener and more sustainable form of construction, despite pushback. Sustainability is important for a variety of reasons, including a better quality of life and environmental quality. In order to have thriving and healthy communities, we need to have clean air, natural resources and a non-toxic environment. With advancements in technology and sustainable materials, the construction industry can lead the way for greener projects. Sustainable construction is developing each and every day, with more demand for cleaner and greener spaces. As the effects of climate change increase, sustainability becomes even more important. While there are challenges, the benefits involved with sustainable construction can create a pathway to a cleaner future.

Rent Equipment for Sustainable Construction Jobs.

Large construction companies aren't the only ones that can change their methods for the betterment of the environment. Regular people working on their own private projects can also focus on utilizing sustainable construction methods. Whether it's using the right equipment, implementing value engineering to determine sustainable material alternatives or simply doing your best to be energy efficient, anyone can help progress sustainability efforts. No matter your next construction project, having the right resources and equipment is the key to a job well done. Big Rentz offers equipment rentals across the United States, making renting machinery easier than ever before. Whether you're taking on a green building project or another construction job, BigRentz has the equipment you need to get the job done.

Eco-Friendly or Green Building Construction systems reduces Environmental impact.

Buildings represent a large part of energy, electricity, water and materials consumption. As of 2020, they account for 37% of global energy use and energy-related CO_2 emissions, which the United Nations estimate contributed to 33% of overall worldwide emissions. Including the manufacturing of building materials, the global CO_2 emissions were 39%. If new technologies in construction are not adopted during this time of rapid growth, emissions could double by 2050, according to the United Nations Environment Program.

Buildings account for a large amount of land. According to the National Resources Inventory, approximately 107 million acres (430,000 km^2) of land in the United States are developed. The International American AgencY released a publication that estimated that existing buildings are responsible for more than 40% of the world's total primary energy consumption and for 24% of global carbon dioxide emissions.

According to Global status report from the year 2016, buildings consume more than 30% of all produced energy. The report states that "Under a below 2°C trajectory, effective action to improve building energy efficiency could limit building final energy demand to just above current levels, meaning that the average energy intensity of the global building stock would decrease by more than 80% by 2050". Green building practices aim to reduce the environmental impact of building as the building sector has the greatest potential to deliver significant cuts in emissions at little or no cost. General guidelines can be summarized as follows: Every building should be as small as possible. Avoid contributing to sprawl. even if the most energy-efficient, environmentally sound methods are used in design and construction. Bioclimatic design principles are able to reduce energy expenditure and by extension, carbon emissions. Bioclimatic design is a method of building design that takes local climate into account to create comfortable conditions within the structure. This could be as simple as constructing a different shape for the building envelope or facing the building towards the south to maximize solar exposure for energy or lighting purposes. Given the limitations of city planned construction, bioclimatic principles may be employed on a lesser scale, however it is still an effective passive method to reduce environmental impact.

What are the Goals of Green Buildings?

Goals of Green buildings:

A green building designed by Michelle Kaufmann. Shanghai Tower, the tallest and largest LEED Platinum certified building in the world since 2015. The concept of sustainable development can be traced to the energy (especially fossil oil) crisis and environmental pollution concerns of the 1960s and 1970s. The Rachel Carson book, "Silent Spring, published in 1962, is considered to be one of the first initial efforts to describe sustainable development as related to green building. The green building movement in the U.S. originated from the need and desire for more Energy-efficient and environmentally friendly construction practices. There are a number of motives for building green, including environmental, economic, and social benefits. However, modern sustainability initiatives call for an integrated and synergistic design to both new construction and in the retrofitting of existing structures. Also known as sustainable design, this approach integrates the building life-cycle with each green practice employed with a design-purpose to create a synergy among the practices used. Green building brings together a vast array of practices, techniques, and skills to reduce and ultimately eliminate the impacts of buildings on the environment and human health. It often emphasizes taking advantage of renewable resources, e.g., using sunlight through passive solar, active solar, and photovoltaic equipment, and using plants and trees through green roofs, rain gardens, and reduction of rainwater run-off. Many other techniques are used, such as using low-impact building materials or using packed gravel or permeable concrete instead of conventional concrete or asphalt to enhance replenishment of groundwater. While the practices or technologies employed in green building are constantly evolving and may differ from region to region, fundamental principles persist from which the method is derived: siting and structure design efficiency, energy efficiency, water efficiency, materials efficiency, indoor environmental quality enhancement, operations and maintenance optimization and waste and toxics reduction. The essence of green building is an optimization of one or more of these principles. Also, with the proper synergistic design, individual green building technologies may work together to produce a greater cumulative effect.

Fig. 7

On the aesthetic side of green architecture or sustainable design is the philosophy of designing a building that is in harmony with the natural features and resources surrounding the site. There are several key steps in designing sustainable buildings: specify 'green' building materials from local sources, reduce loads, optimize systems, and generate on-site renewable energy.

Life Cycle Assessment of Green Buildings.

A life cycle assessment (LCA) can help avoid a narrow outlook on environmental, social and economic concerns by assessing a full range of impacts associated with all cradle-to-grave stages of a process: from extraction of raw materials through materials processing, manufacture, distribution, use, repair and maintenance, and disposal or recycling. Impacts taken into account include (among others) embodied energy, global warming potential, resource use, air pollution, water pollution, and waste.

In terms of green building, the last few years have seen a shift away from a *prescriptive* approach, which assumes that certain prescribed practices are better for the environment, toward the scientific evaluation of actual performance through LCA.

Although LCA is widely recognized as the best way to evaluate the environmental impacts of buildings (ISO 14040 provides a recognized LCA methodology), it is not yet a consistent requirement of green building rating systems and codes, despite the fact that embodied energy and other life cycle impacts are critical to the design of environmentally responsible buildings.

In North America, LCA is rewarded to some extent in the Green Globes rating system, and is part of the new American National Standard based on Green Globes, *ANSI/GBI 01-2010: Green Building Protocol for Commercial Buildings*. LCA is also included as a pilot credit in the LEED system, though a decision has not been made as to whether it will be incorporated fully into the next major revision. The state of California also included LCA as a voluntary measure in its 2010 draft *Green Building Standards Code*.

Although LCA is often perceived as overly complex and time-consuming for regular use by design professionals, research organizations such as BRE in the UK and the Athena Sustainable Materials Institute in North America are working to make it more accessible.

In the UK, the BRE *Green Guide to Specifications* offers ratings for 1,500 building materials based on LCA.

Sitting and structure design efficiency of Green Buildings.

The foundation of any construction project is rooted in the concept and design stages. The concept stage, in fact, is one of the major steps in a project life cycle, as it has the largest impact on cost and performance. In designing environmentally optimal buildings, the objective is to minimize the total environmental impact associated with all life-cycle stages of the building project.

However, building as a process is not as streamlined as an industrial process, and varies from one building to the other, never repeating itself identically. In addition, buildings are much more complex products, composed of a multitude of materials and components each constituting various design variables to be decided at the design stage. A variation of every design variable may affect the environment during all the building's relevant life-cycle stages.

Energy efficiency

Green buildings often include measures to reduce energy consumption – both the embodied energy required to extract, process, transport and install building materials and operating energy to provide services such as heating and power for equipment.

As high-performance buildings use less operating energy, embodied energy has assumed much greater importance – and may make up as much as 30% of the overall life cycle energy consumption. Studies such as the U.S. LCI Database Projec show buildings built primarily with wood will have a lower embodied energy than those built primarily with brick, concrete, or steel.

To reduce operating energy use, designers use details that reduce air leakage through the building envelope (the barrier between conditioned and unconditioned space). They also specify high-performance windows and extra insulation in walls, ceilings, and floors. Another strategy, passive solar building design, is often implemented in low-energy homes. Designers orient windows and walls and place awnings, porches, and trees to shade windows and roofs during the summer while maximizing solar gain in the winter. In addition, effective window placement (daylighting) can provide more natural light and lessen the need for electric lighting during the day. Solar water heating further reduces energy costs.

Onsite generation of renewable energy through solar power, wind power, hydro power, or biomass can significantly reduce the environmental impact of the building. Power generation is generally the most expensive feature to add to a building.

Energy efficiency for green buildings can be evaluated from either numerical or non-numerical methods. These include use of simulation modelling, analytical or statistical tools.

Water efficiency

Reducing water consumption and protecting water quality are key objectives in sustainable building. One critical issue of water consumption is that in many areas, the demands on

the supplying aquifer exceed its ability to replenish itself. To the maximum extent feasible, facilities should increase their dependence on water that is collected, used, purified, and reused on-site. The protection and conservation of water throughout the life of a building may be accomplished by designing for dual plumbing that recycles water in toilet flushing or by using water for washing of the cars. Waste-water may be minimized by utilizing water conserving fixtures such as ultra-low flush toilets and low-flow shower heads. Bidets help eliminate the use of toilet paper, reducing sewer traffic and increasing possibilities of re-using water on-site. Point of use water treatment and heating improves both water quality and energy efficiency while reducing the amount of water in circulation. The use of non-sewage and grey water for on-site use such as site-irrigation will minimize demands on the local aquifer.

Large commercial buildings with water and energy efficiency can qualify for an LEED Certification. Philadelphia's Comcast Center is the tallest building in Philadelphia. It's also one of the tallest buildings in the USA that is LEED Certified. Their environmental engineering consists of a hybrid central chilled water system which cools floor-by-floor with steam instead of water. Burn's Mechanical set-up the entire renovation of the 58 story, 1.4 million square foot sky scraper.

Materials Efficiency

Building materials typically considered 'green' include lumber(that has been certified to a third-party standard), rapidly renewable plant materials (like bamboo and straw), dimension stone, recycled stone, hempcrete, recycled metal *(see: copper sustainability and recyclability)*, and other non-toxic, reusable, renewable, and/or recyclable products. Materials with lower embodied energy can be used in substitution to common building materials with high degrees of energy consumption and carbon/harmful emissions. For concrete a high performance self-healing version is available, however options with lower yields of pollutive waste entertain ideas of upcycling and congregate supplementing; replacing traditional concrete mixes with slag, production waste, and aggregates. Insulation also sees multiple angles for substitution. Commonly used fiberglass has competition from other eco-friendly, low energy embodying insulators with similar or higher R-values (per inch of thickness) at a competitive price. Sheep wool, cellulose, and Therma Cork perform more efficiently, however, use may be limited by transportation or installation costs.

Furthermore, embodied energy comparisons can help deduce the selection of building material and its efficiency. Wood production emits less CO_2 than concrete and steel if produced in a sustainable way just as steel can be produced more sustainably through improvements in technology (e.g. EAF) and energy recycling/carbon capture(an underutilized potential for systematically storing carbon in the built environment).

The EPA (Environmental Protection Agency) also suggests using recycled industrial goods, such as coal combustion products, foundry sand, and demolition debris in construction projects. Energy efficient building materials and appliances are promoted in the United States through energy rebate programs.

A 2022 report from the Boston Consulting Group found that, investments in developing greener forms of cement, iron, and steel lead to bigger greenhouse gas reductions compared with investments in electricity and aviation. In addition, the process of making cement without producing CO_2

is unavoidable. However, using pozzolans clinkers can reduce CO_2 emission while in the process of making cement.

Deconstruction of Buildings.

In the context of physical construction, deconstruction is the selective dismantlement of building components, specifically for Reuse, repurposing, recycling and waste management. It differs from Demolition, where a site is cleared of its building by the most expedient means. Deconstruction has also been defined as "construction in reverse". Deconstruction requires a substantially higher degree of hands-on labor than does traditional demolition, but as such provides a viable platform for unskilled or unemployed workers to receive job skills training. The process of dismantling structures is an ancient activity that has been revived by the growing fields of sustainable and green building.

When buildings reach the end of their useful life, they are typically demolished and hauled to Landfills, Building implosions or 'wrecking-ball' style Demolition are relatively inexpensive and offer a quick method of clearing sites for new structures. On the other hand, these methods create substantial amounts of waste. Components within old buildings may still be valuable, sometimes more valuable than at the time the building was constructed. Deconstruction is a method of harvesting what is commonly considered "waste " and reclaiming it in to useful building material. Most modern buildings are difficult to perform deconstruction due to the designs of such buildings.

Contribution to sustainability

Deconstruction has strong ties to environmental sustainability. In addition to giving materials a new life cycle, deconstructing buildings helps to lower the need for virgin resources. This in turn leads to energy and emissions reductions from the refining and manufacture of new materials, especially when considering that an estimated 40% of global material flows can be attributed to construction, maintenance, and renovation of structures. As deconstruction is often done on a local level, many times on-site, energy and emissions are also saved in the transportation of materials. Deconstruction can potentially support communities by providing local jobs and renovated structures. Deconstruction creates 6-8 jobs, for every job created by traditional demolition. In addition, solid waste from conventional demolition is diverted from landfills. This is a major benefit because Construction and demolition waste accounts for approximately 20% - 40% of the solid waste stream. 90% of this construction and demolition waste stream is generated during the process of demolition. In 2015 548 million tons of Construction and demolition waste were created in the United States alone.

Deconstruction allows for substantially higher levels of material reuse and recycling than does conventional processes of demolition. Up to 25% of material in a traditional residential structure can be readily reused, while up to 70% of material can be recycled.

Typical Methods of Deconstruction.

1. Deconstruction is commonly separated into two categories; Structural and non-structural. Non-structural deconstruction, also known as "soft-stripping", consists

   of reclaiming non-structural components, appliances, doors, windows , and finish materials.

2. The reuse of these types of materials is commonplace and considered to be a mature market in many locales.

3. Structural deconstruction involves dismantling the structural components of a building. Traditionally this had only been performed to reclaim expensive or rare materials such as used brick, dimension stone, and extinct wood . In antiquity, it was common to raze stone buildings and reuse the stone; it was also common to steal stones from a building that was not being totally demolished: this is the literal meaning of the word. Used brick and dimension limestone, in particular, have a long tradition of reuse due to their durability and color changes over time. Recently, the rise of environmental awareness and sustainable building has made a much wider range of materials worthy of structural deconstruction.[13] Low-end, commonplace materials such as dimensional number have become part of this newly emerging market.

4. The united states military has utilized structural deconstruction in many of its bases. The construction methods of barracks , among other base structures, are usually relatively simple. They typically contained large amounts of lumber and used minimal adhesives and finish-work. In addition, the buildings are often identical, making the process of deconstructing multiple buildings much easier. Many barracks were the era, and have aged to the point where they now need to be torn down. Deconstruction was deemed very practical due to the abundance of labor the military has access to and the value of the materials themselves.

5. Natural disaster , such as hurricanes,floods.tsunamis , and earthquake often leave a vast amount of usable building materials in their wake. Structures that remain standing are often deconstructed to provide materials for rebuilding the region.

Economic Potential

6. Deconstruction's economic viability varies from project to project. The amount of time and cost of labor are the main drawbacks. Harvesting materials from a structure can take weeks, whereas demolition may be completed in roughly a day. However, some of the costs, if not all, can be recovered. Reusing the materials in a new on-site structure, selling reclaimed materials, donating materials for Income tax write-offs, and avoiding landfill "tipping fees" are all ways in which the cost of deconstruction can be made comparable to demolition.

7. Reclaiming the materials for a new on-site structure is the most economically and environmentally efficient option. Tipping fees and the costs of new materials are avoided; in addition, the transportation of the materials is non-existent. Selling the used materials or donating them to non-profit organizations are another effective way of gaining capital. Donations to NPO's such as habitat for humanity's ReStore are tax-deductible. Many times it is possible to claim the value to be half of what that particular material would cost new. When donating rare or antique components it is sometimes possible to claim a higher value than a comparable, brand-new material.

8. Value can also be added to new structures that are built by implementing reused materials. The United States Green Building Council's program entitled Leadership in Energy and Environmental Design (LEED) offers seven credits relating to reusing materials. (This accounts for seven out of a maximum sixty-nine credits) These include credits for building-shell reuse, material reuse, and diverting waste from landfills. Building shell-reuse is particularly appropriate for shells made of dimension stone.

9. Deconstruction is well suited to job training for the construction trades. Taking down a building is an excellent way for a worker to learn how to put a building up. This is vital for the economic recovery of inner-city communities. Unskilled and low-skilled workers can receive on-the-job training in use of basic carpentry tools and techniques, as well as learning teamwork, problem-solving, critical thinking and good work habits.

Deconstruction vs. Demolition.

As opposed to the method of demolishing a building, deconstructing a building is a much safer method for both the environment, as well as the overall health of humans in terms of air pollution. Structures are usually taken down using the method of implosion, where explosives are used to implode the building on itself. This in itself causes a variety of harmful substances to enter the atmosphere and affect our air quality. Although not done through the method of implosion, the September 11, 2001 attacks on New York City's World Trade Center serve as a good reference point to the harmful effects that come along with the demolition of such large structures such as these buildings. The reason for this is primarily because of the similarities between a controlled demolition and the way in which the Twin Towers collapsed that day. The environmental effects that followed these attacks included the release of numerous harmful and toxic particles into the air, which had a huge impact on New York City's air quality. Not only has this been detrimental to the environment, but also to the physical health of many people. In many instances, the substances that are released by these practices are directly linked to numerous diseases and illnesses found in many people who have been within a certain proximity of a demolition. Again relating to 9/11, there have been countless instances and studies performed to show how these ailments arose in 9/11 survivors. As a healthier alternative, deconstruction is used in many instances due to the fact that it does not share any of the same negative affects to the air quality with its counter method of demolition. As previously stated above, this method involves carefully taking the building apart through the dismantling of each part, ultimately reducing the amount of pollutants released into the environment, as well as aiding in the processes of recycling and waste management. It is because of this that so many believe deconstruction to be a much safer and environmentally-friendly method of taking down structures.

An upstream approach to deconstruction can be implemented into buildings during their design process, known as designing for deconstruction (DfD). This is a current trend in Sustainable architecture. DFD structures typically use simple construction methods combined with high-grade, durable materials. Separating layers of a building's infrastructure and making them visible can significantly simplify its deconstruction. Making components within systems separable also assists in being able to dismantle materials quickly and efficiently. This can be achieved by using mechanical fasteners such as bolts to connect parts. Allowing physical access to these fasteners is another necessary aspect of this design, as well as the use of standardized materials assembled consistently throughout the project. Consolidation of plumbing, HVAC, and other utility service points within a building reduces the need for long service lines, as well as points of entanglement and conflict with other building elements. Similarly, utilizing raised floor or dropped ceiling methods allows easier access to mechanical and electrical services, and can reduce the time needed to remove these components during the process of deconstruction. Some conventional construction methods and materials are difficult or impossible to deconstruct; the use of nails and adhesives significantly slows down the deconstruction process and can render unusable materials that could otherwise be reused. The presence of hazardous materials is also an obstacle for deconstruction. Using mixed material grades makes the process of identifying pieces for resale difficult.

Some commercial buildings that have been designed according to DFD principles use built in anchor points and other features intended to provide additional fall protection options. Such design considerations can increase overall worker safety, and decrease amount of overall time spent on deconstruction.

DFD not only enables the end of a building's life-cycle, but can also make the building easier to maintain and adapt to new uses. Saving the shell of a building or adapting interior spaces to meet new needs can reduce the environmental impact of new structures.

Modular Construction
Advantage
Time
Production algorithms
Cost
Transportation
Construction
Safety.
Sustainability.

Modular Construction Systems.

Modular construction is a construction technique which involves the prefabrication of 2D panels or 3D volumetric structures in off-site factories and transportation to construction sites for assembly. This process has the potential to be superior to traditional building in terms of both time and costs, with claimed time savings of between 20 and 50 percent faster than traditional building techniques.

It is estimated that by 2030, modular construction could deliver US$22 billion in annual cost savings for the US and European construction industry, helping fill the US$1.6 trillion productivity gap. The current need for standardized, repeatable 3D volumetric housing pre-fabricated units and designs for student accommodations, affordable housing and hotels is driving demand for modular construction.

A. Advantages:

In a 2018 *Practice Note*, the states that the benefits obtained from offsite construction mainly relate to the creation of components in a factory setting, protected from the weather and using manufacturing techniques such as assembly lines with dedicated and specialist equipment.[2] Through the use of appropriate technology, modular construction can:

- increase the speed of construction by increasing the speed of manufacture of the component parts,
- reduce waste,
- increase economies of scale,
- improve quality leading to reduction in the whole life costs of assets
- reduce environmental impact such as dust and noise and
- reduce accidents and ill health by reducing the amount of construction work taking place at site.

B. Time:

Modular construction has consistently been at least 20 percent faster than traditional on-site builds. Currently, the design process of modular construction projects tends to take longer than that of traditional building. This is because modular construction is a fairly new technology and not many architects and engineers have experience working with it. In fewer words, the industry has not yet learned how to work this way. It is expected of design firms to develop module libraries which would assist in the automatization of this process. These modules libraries would hold various pre-designed 2D panels and 3D structures which would be digitally assembled to create standardized structures.

The foundations of a structure are a crucial part of its rigidity. The magnitude and complexity of such will vary depending on the size, and overall weight of the structure. Therefore, the weight difference of a traditionally built house and a prefabricated structure will mean that foundations needed will be smaller and faster to build.

Off-site manufacturing is the pinnacle of modular construction. The ability to coordinate and repeat activities in a factory along with the increased help of automation result in largely faster manufacturing times than those of on-site building. A large time saver is the ability to parallelly work on the foundation of a structure and the manufacturing of the structure itself. This would be impossible with traditional construction. The on-site construction is radically simplified. The assembly of pre-fabricated components is as simple as assembling the 3D modules, and connecting the services to main site connections. A team of five workers can assemble up to six 3D modules, or the equivalent of 270 square meters of finished floor area, in a single work day.

C. Production Algorithms:

Since the technology required to manufacture the components of modular construction, the prefabricated parts of modular buildings are carried out by modular factories. To optimize time, modular factories consider the specifications and resources of the project and adapt a scheduling algorithm to fulfill the needs of this unique project. However, current scheduling methods assume the quantity of resources will never reach zero, therefore representing an unrealistic work cycle.

A modular factory handling a single project at any given point is rare, and would produce low returns. Hyun and Lee's research propose a Genetic Algorithm (GA) scheduling model which takes into consideration various project's characteristics and shares resources. The production sequence of this algorithm would be largely affected by which modules need to be transported to which site and the dates they should arrive. After considering the variables of production, transportation and on-site assembly the objective function is:

Where S_i is the number of stocked units per day, P_i is the number of units per day and E_i is number of units installed per day. Production algorithms are continuously being developed to further accelerate the production of modular construction buildings, enlarging the time saving gap with traditional construction methods.

D. Cost:

Modular construction can yield up to 20 percent of the total project cost in savings. However, there is also a risk of it increasing the cost by 10 percent. This occurs when the savings in the labor area of construction are outweighed by the increase in costs of the logistics area and materials. The pre-fabrication of components used in modular construction have a higher logistics cost than traditional building. Since the panels or 3D structures have to be manufactured in a factory and transported to the construction site, new variables which alter the flow of construction are presented.

E. Transportation:

Transportation of fabricated components is naturally more expensive than that of raw materials. For one, even a number of 2D panels stacked together are much harder to transport than the raw cement, wood or material used to build them. Panels run a high risk of suffering minor or major damage when being transported through land. If a panel were to be damaged, it would likely have to be replaced entirely. The factory would need to temporarily stop production of other panels to replace this one, increasing the overall manufacturing hours and therefore cost. On top of the manufacturing hours, the transportation hours would also be increased, increasing yet another cost. Regardless, the transportation of 2D panels is still a good alternative to on-site construction.

Transportation reaches its peak cost when shipping 3D volumetric structures. While 1 m^2 of 2D floor space takes approximately US$8 to transport 250 km, its equivalent in 3D floor space takes US$45.[1] Adding to this the replacement cost if the structure gets damaged during transport creates a large cost increase.

F. Construction:

Assembling components in a factory off-site means that workers can use the repeatability of the structures as well as the use of automation to facilitate the manufacturing process. By standardizing the overall design of structures, work which would usually require expensive workers with specific skills (e.g. mechanical, electrical and plumbing) can be completed by low-cost manufacturers, decreasing the total salaries cost. As very little manufacturing occurs on-site, up to 80% of traditional labor activity can be moved off-site to the module factory. This leads to a lower number of sub-contractors needed, further decreasing overall total salaries cost. Overall, the larger the labor-intensive portion of a project, the larger the savings will be if modular construction is used.

Project such as student accommodations, hotels and affordable housing are great candidates for modular construction. The repeatability of their structures leads to faster manufacturing times and therefore less overall cost. Meanwhile, if the project is (for example) a modern beach house with highly irregular wall spaces and ceilings, traditional construction methods may be preferable. As the industry continues to adapt and grow, these repeatable designs could one day be modified and adapted to fit all kinds of structures at decreased costs.

G. Safety:

Construction is considered to be one of the most dangerous industries. Workers fall from heights, objects are dropped, muscles are strained and environmental hazards can be found. Modular construction constrains all manufacturing activities to a ground level, clean space with fewer workers needed. It is estimated that reportable accidents are reduced by over 80% relative to site-intensive construction.[4] When asked in a survey about safety management in the construction industry conducted by McGraw Hill Construction in 2013, 50% of the construction industry believed that pre-fabrication was safer than traditional on-site building, while only 4% said that prefabrication or modular construction had a negative impact on safety performance. Of the general and specialty contractors surveyed, 78% and 59% said that the largest safety impact was the undergoing of complex tasks at ground level.[5] According to the CDC, falling is the leading cause of work-related fatalities in construction, making up more than one in every three deaths in the industry.[6] The reduction of heights at which workers need perform tasks on subsequently reduces the fatality risk they experience, greatly increasing the overall safety of the industry. Also, 69% of the general contractors as well as 69% of the specialty contractors mentioned that the reduced number of workers performing different tasks at the off-site factory also improved construction site safety. Overall, modular construction is safer for the following reasons:

- Stable work location
- Tasks are performed in ample spaces
- Ground level assembly
- Cover from harsh weather
- Better monitoring of unsafe activities
- 30 to 50 percent reduction in time spent at construction site
- Fewer personnel on-site

Modular construction is still not considered an entirely safe alternative. However, it does reduce accidents and fatalities by a significant amount.

Especially in the manufacturing process of a project. 48.1% of all accidents during on-site construction were fall-related, while only 9.1% of the accidents at manufacturing plants were from falls.[5] Manufacturing plant workers were more likely to be struck by an object or equipment (37.1%) and fracture and amputation had the same injury type frequency at 27.3%. Nevertheless, as the construction industry continues to adapt and moves over to more sustainable construction methods like pre-fabricated modular construction, it is expected that the overall safety number of accidents at construction sites will decrease.

The use of modular construction methods is encouraged by proponents of Prevention through Design techniques in construction. It is included as a recommended hazard control for construction projects in the "PtD - Architectural Design and Construction Education Module" published by the National Institute for Occupational Safety and Health.

H. Sustainability:

Modular construction is a great alternative to traditional construction when looking at the amount of waste each method produces. When constructing a high-rise building in Wolverhampton, 824 modules were used. During this process about 5% of the total weight of the construction was produced as waste. If it is compared to traditional methods' 10-13% average waste a small difference can be observed.[4] This difference may not seem like much when talking about small structures, however when talking about a 100,000 lb/ft^2 building it is a significant percentage. Also, the number of on-site deliveries decreased by up to 70%.[4] The deliveries are instead moved to the modular factory, where more material can be received. On-site Noise pollution is greatly reduced as well, by moving the manufacturing process to an off-site factory, usually located outside of the city, neighboring buildings are not impacted as they would with the traditional building process.

I. Modular Construction Systems:

Open-source and commercial hardware components used in modular construction include: open beams, bit beams, maker beams, grid beams, contraptors, Open Structures components, etc. Space frame systems (such as Mero, Unistrut, Delta Structures, etc.) also tend to be modular in design.

Other materials used in construction which are interlocking and thus reusable/modular in nature include interlocking bricks.

Explanation about Prefabricated Buildings.

It is to be defined that to manufacture the parts of something beforehand so that it can be built by putting the parts together. Prefabrication is the practice of assembling components of a structure off-site in a warehouse or other location and transporting the complete assemblies to the construction site.

History of Prefabricated Buildings.

Buildings have been built in one place and reassembled in another throughout history. This was especially true for mobile activities, or for new settlements. Elmina Castle, the first slave fort in West Africa, was also the first European prefabricated building in Sub-

saharan Africa. In North America, in 1624 one of the first buildings at Cape Ann was probably partially prefabricated, and was rapidly disassembled and moved at least once. John Rollo described in 1801 earlier use of portable hospital buildings in the West Indies. Possibly the first advertised prefab house was the "Manning cottage". A London carpenter, Henry Manning, constructed a house that was built in components, then shipped and assembled by British emigrants. This was published at the time (advertisement, South Australian Record, 1837) and a few still stand in Australia.[3] One such is the Friends Meeting House, Adelaide. The peak year for the importation of portable buildings to Australia was 1853, when several hundred arrived. These have been identified as coming from Liverpooll, Boston and Singapore (with Chinese instructions for re-assembly). In Barbados the Chattel house was a form of prefabricated building which was developed by emancipated slaves who had limited rights to build upon land they did not own. As the buildings were moveable they were legally regarded as chattels.

Fig. - 8

Buildings have been built in one place and reassembled in another throughout history. This was especially true for mobile activities, or for new settlements. Elmina Castle, the first slave fort in West Africa, was also the first European prefabricated building in Sub-saharan Africa. In North America, in 1624 one of the first buildings at Cape Ann was probably partially prefabricated, and was rapidly disassembled and moved at least once. John Rollo described in 1801 earlier use of portable hospital buildings in the West Indies. Possibly the first advertised prefab house was the "Manning cottage". A London carpenter, Henry Manning, constructed a house that was built in components, then shipped and assembled by British emigrants. This was published at the time (advertisement, South Australian Record, 1837) and a few still stand in Australia.[3] One such is the Friends Meeting House, Adelaide. The peak year for the importation of portable buildings to Australia was 1853, when several hundred arrived. These have been identified as coming from Liverpooll, Boston and Singapore (with Chinese instructions for re-assembly). In Barbados the Chattel house was a form of prefabricated building which was developed by emancipated slaves who had limited rights to build upon land they did not own. As the buildings were moveable they were legally regarded as chattels.

In 1855 during the Crimean War, after Florence Nightingale wrote a letter to *The Times*, Isambard Kingdom Brunel was commissioned to design a prefabricated modular hospital. In five months he designed the Renkioi Hospital: a 1,000 patient hospital, with innovations in sanitation, ventilation and a flushing toilet. Fabricator William Eassie constructed the required 16 units in Gloucester Docks, shipped directly to the Dardanelles.

Only used from March 1856 to September 1857, it reduced the death rate from 42% to 3.5%.

The world's first prefabricated, pre-cast panelled apartment blocks were pioneered in Liverpool. A process was invented by city engineer John Alexander Brodie, whose inventive genius also had him inventing the football goal net. The tram stables at Walton in Liverpool followed in 1906. The idea was not extensively adopted in Britain, however was widely adopted elsewhere, particularly in Eastern Europe.

Prefabricated homes were produced during the Gold Rush in the United States, when kits were produced to enable Californian prospectors to quickly construct accommodation. Homes were available in kit form by mail order in the United States in 1908.[9]

Prefabricated housing was popular during the Second World War due to the need for mass accommodation for military personnel. The United States used Quonset huts as military buildings, and in the United Kingdom prefabricated buildings used included Nissen huts and Bellman Hangars. 'Prefabs' were built after the war as a means of quickly and cheaply providing quality housing as a replacement for the housing destroyed during the Blitz. The proliferation of prefabricated housing across the country was a result of the Burt Committee and the Housing (Temporary Accommodation) Act 1944. Under the Ministry of Works Emergency Factory Made housing programme, a specification was drawn up and bid on by various private construction and manufacturing companies. After approval by the MoW, companies could bid on Council led development schemes, resulting in whole estates of prefabs constructed to provide accommodation for those made homeless by the War and ongoing slum clearance. Almost 160,000 had been built in the UK by 1948 at a cost of close to £216 million. The largest single prefab estate in Britain was at Belle Vale (South Liverpool), where more than 1,100 were built after World War

2. The estate was demolished in the 1960s amid much controversy as the prefabs were very popular with residents at the time.

Prefabs were aimed at families, and typically had an entrance hall, two bedrooms (parents and children), a bathroom (a room with a bath) — which was a novel innovation for many Britons at that time, a separate toilet, a living room and an equipped (not *fitted* in the modern sense) kitchen. Construction materials included steel, aluminium, timber or asbestos, depending on the type of dwelling. The aluminium *Type B2* prefab was produced as four pre-assembled sections which could be transported by lorry anywhere in the country.

The Universal House (pictured left & lounge diner right) was given to the Chiltern Open Air Museum after 40 years temporary use. The Mark 3 was manufactured by the Universal Housing Company Ltd, Rickmansworth.

The United States used prefabricated housing for troops during the war and for GIs returning home. *Prefab* classrooms were popular with UK schools increasing their rolls during the baby boom of the 1950s and 1960s.

Many buildings were designed with a five-ten year life span, but have far exceeded this, with a number surviving today. In 2002, for example, the city of Bristol still had residents living in 700 examples. Many UK councils have been in the process of demolishing the last surviving examples of Second World War prefabs in order to comply with the British government's Decent Homes Standard, which came into effect in 2010. There has, however, been a recent revival in prefabricated methods of construction in order to compensate for the United Kingdom's current housing shortage.

Explain about Prefabricated and Modernist Movement.

Architects are incorporating modern designs into the prefabricated houses of today. Prefab housing should no longer be compared to a mobile home in terms of appearance, but to that of a complex modernist design. There has also been an increase in the use of "green" materials in the construction of these prefab houses. Consumers can easily select between different environmentally friendly finishes and wall systems. Since these homes are built in parts, it is easy for a home owner to add additional rooms or even solar panels to the roofs. Many prefab houses can be customized to the client's specific location and climate, making prefab homes much more flexible and modern than before.

There is a zeitgeist or trend in architectural circles and the spirit of the age favors the small carbon footprint of "prefab".

Efficiency

The process of building pre-fabricated buildings has become so efficient in China that a builder in Changsha built a ten-storey building in 28 hours and 45 minutes.

Sustainability

Prefabricated construction generates less carbon footprint, improves energy use and efficiency, and produces less waste, making it more sustainable and environmentally friendly, and compliant with sustainable design standards.

Modular Architecture

The modular architecture allows, thanks to 3D modeling, the design and construction of the modular structure outside the site where it will be installed. [19] This offers several advantages such as more sustainable design, greater cost and time savings and standardization of design. [20] This is especially important for large-scale construction projects.

Prefabricated Commercial Buildings.

Buildings have been built in one place and reassembled in another throughout history. This was especially true for mobile activities, or for new settlements. Elmina Castle, the first slave fort in West Africa, was also the first European prefabricated building in Sub-saharan Africa, In North America, in 1624 one of the first buildings at Cape Ann was probably partially prefabricated, and was rapidly disassembled and moved at least once. John Rollo described in 1801 earlier use of portable hospital buildings in the West Indies. Possibly the first advertised prefab house was the "Manning cottage". A London carpenter, Henry Manning, constructed a house that was built in components, then shipped and assembled by British emigrants. This was published at the time (advertisement, South Australian Record, 1837) and a few still stand in Australia. One such is the Friends Meeting House, Adelaide. The peak year for the importation of portable buildings to Australia was 1853, when several hundred arrived. These have been identified as coming from Liverpool, Boston and Singapore (with Chinese instructions for re-assembly). In Barbados the Chattel house was a form of prefabricated building which was developed by emancipated slaves who had limited rights to build upon land they did not own. As the buildings were moveable they were legally regarded as chattels. In 1855 during the Crimean War, after Florence Nightingale wrote a letter to *The Times*, Isambard Kingdom Brunel was commissioned to design a prefabricated modular hospital. In five months he designed the Renkioi Hospita: a 1,000 patient hospital, with innovations in sanitation, ventilation and a flushing toilet. Fabricator William Eassie constructed the required 16 units in Gloucester Docks, shipped directly to the Dardanelles. Only used from March 1856 to September 1857, it reduced the death rate from 42% to 3.5%.

The world's first prefabricated, pre-cast panelled apartment blocks were pioneered in Liverpol. A process was invented by city engineer John Alexander Brodie, whose inventive genius also had him inventing the football goal net. The tram stables at Walton in Liverpool followed in 1906. The idea was not extensively adopted in Britain, however was widely adopted elsewhere, particularly in Eastern Europe.

Prefabricated homes were produced during the Gold Rush in the United States, when kits were produced to enable Californian prospectors to quickly construct accommodation. Homes were available in kit form by mail order in the United States in 1908.

Prefabricated housing was popular during the Second World War due to the need for mass accommodation for military personnel. The United States used Quonset huts as military buildings, and in the United Kingdom prefabricated buildings used included Nissen huts and Bellman Hangars. 'Prefabs' were built after the war as a means of quickly and cheaply providing quality housing as a replacement for the housing destroyed during the Blitz. The proliferation of prefabricated housing across the country was a result of the Burt Committee and the Housing (Temporary Accommodation) Act 1944. Under the Ministry of Works Emergency Factory Made housing programme, a specification was drawn up and bid on by various private construction and manufacturing companies. After approval by the MoW, companies could bid on Council led development schemes, resulting in whole estates of prefabs constructed to provide accommodation for those made homeless by the War and ongoing slum clearance. Almost 160,000 had been built in the UK by 1948 at a cost of close to £216 million. The largest single prefab estate in Britain was at Belle Vale (South Liverpool), where more than 1,100 were built after World War 2. The estate was

demolished in the 1960s amid much controversy as the prefabs were very popular with residents at the time.

Prefabs were aimed at families, and typically had an entrance hall, two bedrooms (parents and children), a bathroom (a room with a bath) — which was a novel innovation for many Britons at that time, a separate toilet, a living room and an equipped (not *fitted* in the modern sense) kitchen. Construction materials included steel, aluminium, timber or Asbestos, depending on the type of dwelling. The aluminium *Type B2* prefab was produced as four pre-assembled sections which could be transported by lorry anywhere in the country.

What Are the Benefits of Eco-Friendly Buildings?

Broadly, the benefits of eco-friendly buildings can be divided into three main categories: Environmental, Economic, and social benefits.

The benefits within each category overlap to some extent and different green buildings may provide more benefits in one area than another. However, all Eco-friendly buildings offer at least some benefit in all three areas.

Let's look at some of the key aspects within each category of benefits:

Environmental Benefits of Eco-Friendly Buildings

Eco-friendly buildings are built with the natural environment in mind. They are designed and constructed to conserve natural resources, produce less waste and emissions, and utilize sustainable materials, that have been ethically sourced.

In some cases, they can even result in positive impacts for the environment through carbon capture, providing habitat for flora and fauna, and promoting local ecological health. They're also able to provide natural resources to the local area, such as solar power to supplement local energy needs. The environmental benefits of eco-friendly buildings contribute to slowing down climate change, reducing greenhouse gas and carbon emissions.

Buildings that meet green building requirements, such as LEED and Green Building Council standards, use fewer resources by between 11% and 60% when compared to minimum industry standards, depending on the certification requirements of their local 'green authority'

Economic Benefits of Eco-Friendly Buildings

There are many economic benefits of eco-friendly buildings. These benefits include:

- Reduced utility bills for tenants and owners as electricity and water resources are used more efficiently and supplemented with renewable energy.
- Lower construction costs and higher property values
- Higher occupancy rates and increased productivity for businesses
- Lower operational and maintenance costs
- Contribute to the city and countrywide savings on electricity and water, as well as waste management
- Job creation in growing sectors that deal specifically with green or sustainable construction and maintenance
- Contribute to growing green industries that manufacture materials and technology for renewable energy and other sustainable materials
- Contribute to local economies when locally sourced materials are used to reduce environmental impacts from the manufacture and transportation of materials
- Tax concessions and financial incentives boost development in areas where these are offered, which boosts local economies and in

turn reduces the burden on the state to meet the demands of a growing local economy and greater development.

1. Social benefits of Eco-Friendly Building.

Eco-friendly buildings contribute to positive social impacts for their occupants, neighbors, and broader communities. These benefits include:

- Better health and well-being for people who live and work in eco-friendly buildings
- Increased cognitive scores and productivity for people working in green, well-ventilated spaces
- Better sleep associated with reduced stress levels and improved mood for those living and working in buildings designed to be green and beneficial to human health and well-being.

Eco-Friendly Buildings to See Around the World.

1. The Edge – Amsterdam, Netherlands

Located in Amsterdam, Netherlands, The Edge is an office block that features a large glass atrium and expansive open spaces, walled with windows. The building is a smart building, operated from an app. It is focused on being a green building and on occupant comfort and productivity.

Lighting and power use reduction are at the core of The Edge's green credentials. Power to the building is from the solar panels that cover the exterior. Internal climate is controlled through a system that pumps warmer and cooler water from an aquifer to different levels in the building.

In place of traditional lights, the building utilizes a combination of natural light and sensor-

In place of traditional lights, the building utilizes a combination of natural light and sensor-controlled LEDs that adjust their brightness as needed. This helps the building save up to 80% of the power it would require if it used traditional electric lights running at a steady level constantly.

Natural lighting can be controlled through the app, which allows occupants to open or close their window blinds as needed.

The innovative, smart design and their impressive reduction of power usage helped The Edge achieve a BREEAM score of 98.3%!

2. Acros Fukuoka Prefectural International Hall – Fukuoka, Japan

Situated in Fukuoka, Japan, the Acros Fukuoka Prefectural International Hall is a mixed-use building that houses government offices, exhibition centers, and a concert hall.

The building is a synthesis of functional building and open, green space. One side of the building is a commercial building and the other is an impressive, terraced garden that doubles as recreational spaces for the inhabitants of Fukuoka to enjoy.

The terraced gardens help to regulate building temperatures and reduce power consumption while providing valuable habitat for local fauna within an otherwise densely developed urban area.

3. Bosco Verticale – Milan, Italy

The Bosco Verticale towers in Milan, Italy, are a pair of residential towers that house a combination of human tenants and a huge variety of plant and animal species.

The luxury apartment blocks were designed to house natural plant life, including fully grown trees, that are the equivalent of one hectare of natural forest.

The building also uses innovative geothermic heat pump technology to regulate temperatures and reduce power consumption. Power is supplemented through solar panels on the building.

With Gold LEED Certification, the 'vertical forest' provides valuable habitat and has created a thriving ecosystem within the city. The building and its lush vegetation serve to capture carbon (19.000 kg/year) and produce oxygen (18.980 kg/year) within a metropolitan area. This contributes positively to climate change, as well as reducing air pollution.

4. Vancouver Convention Centre West – Vancouver, Canada

The west building of the Vancouver Convention Centre in Vancouver, Canada, was the first green building to achieve double Platinum LEED certification.

For its human occupants and visitors, the building serves as a combination of convention center spaces, retail outlets, art galleries, and numerous public open spaces. The building makes use of solar power to reduce energy consumption and also utilizes seawater pumped through the buiA 6-acre rooftop garden covers the roof, providing habitat for local flora and fauna, as well as four colonies of bees. The roof slopes to facilitate drainage, water purification, and preservation, and the spread of seeds from the gardens.

Below, the shorefront portion of the building is built on columns in the sea, which support five tiers at different depths for a variety of marine habitats.

This has created a complete shoreline ecosystem that includes tens of different species, such as salmon, crabs, starfish, and shellfish. In addition to the fully submerged habitats, there are tidal habitats created by runnels that flush and drain daily beneath the building.

The Vancouver Convention Centre is a great example of how we can design and operate buildings that seamlessly integrate human and ecological interests to the best benefit of both.

5. <u>One Central Park (Sydney, Australia)</u>

One Central Park is a mixed-use residential and retail building in Sydney, Australia. The two residential towers and six-level shopping center are positioned on the edge of a park, which doesn't end where the buildings begin!

One Central Park hosts hanging gardens that extend up the façade of the towers and rooftop gardens on top of the lower levels.

The building is largely self-sufficient, using a low carbon tri-generation power plant and an internal membrane bioreactor water recycling plant. This, and their contributions to reducing emissions and air pollution earned One Central Park a 5 Star Award Certification from the Green Building Council of Australia.

Eco-Friendly Solutions to Lower Your Utility Bill.

This may seem gloomy, but the good news is that individuals do have the power to make greener choices that reduce their reliance on fossil fuels and cut back on energy consumption while utility prices remain higher than average. You can try eco-friendly solutions to focus your efforts on controllable things, like personal electricity consumption. Eradicating high energy bills might be out of reach, but you can do your part to cut corners and live sustainably where you can.

1. Unplug Power-Drawing Devices

Did you know that electronic devices and appliances you have plugged into your outlets can continually uses power even when they aren't in use? Unplug electronics and appliances when you don't need them to cut back on your energy consumption.

Want to see the effects this while have before you try it? You can take time to test appliances and devices in your home to see if they are drawing unnecessary energy when not in use. Use a watt-meter to guage whether devices or appliances are registering power when turned off. If so, you know they can be unplugged or upgraded to a more sustainable model. You could also plug everything into power strips with a switch. This is great for computer setups and entertainment consoles. Toggle the switch off when you aren't using the products so it conserves power.

2. Rework Your Routine to Avoid Peak Demand

In the summer and in hot climates, appliances use the most energy during the hottest parts of the day. Cooking on a stove at high noon or running a washing machine at peak hours will make your home event hotter, causing your AC to do more work to compensate.

In winter and in colder climates, this peak demand period is often a bigger concern. With all homeowners on the grid pulling more power for heating during winter freezes, fuel costs rise and energy consumption becomes more expensive.

What can you do to help? A change in routine can help you conserve this energy without any fancy upgrades or maintenance. When the heat is scorching midday or a snowstorm is hitting your home, consider waiting to do laundry or run the dishwasher.

Instead, run your dishwasher in the early morning hours before the heat of the day arrives and do your laundry later in the evening. Little changes to your daily schedule can affect your electricity bill over time — and tweaking your house cleaning routine might help you feel more productive to boot.

3. Energy Efficient Appliances

Ready for an upgrade? Purchase energy-efficient appliances that can help save electricity. The Energy Star program certifies models that consume less energy and support environmental goals. Energy Star rated appliances are 10-25% more efficient than standard options. There are smaller upgrades you can make, too. Low flow water fixtures are also an asset to lower utility costs. Low-flow showerheads and bath spigots reduce both water and energy consumption. Your appliances are one of the biggest sources of energy consumption in your home. Make some eco-friendly switches and you'll see positive changes on your utility bill while making a positive impact on the environment.

4. LED Lights

Your lights are an even easier upgrade for energy efficiency. LED lightbulbs are the easiest and most cost-effective switch to cut costs. LED lightbulbs are long-lasting and draw much less energy than their counterparts. They're 75% more efficient than incandescent bulbs and last years longer, too.

5. HVAC Check

Your HVAC system routinely cools your home in the summer and heats it in the winter. You should change the air filters regularly and check your HVAC regularly for any faults or issues. These systems require maintenance to run at optimal capacity and distribute clean air throughout your home.

When leaks are undetected or your thermostat has faults, the system will overcompensate and use more electricity than necessary. To save on energy consumption, conduct regular checks and keep your filters rotating. Hire a professional to do regular inspections to reduce energy use continually.

Some lifestyle adjustments can also lessen the impact HVAC has on your utility bills each month. While there's no need to sacrifice your health and comfort, learning to live with temperatures a few degrees warmer or cooler throughout different seasons can add up over time. Make sure the clothes you wear are weather-appropriate and you adjust the thermostat for extended absences to reduce HVAC uses.

6. Long-Term Switch: Renewable Energy

The most impactful way to reduce your utility bills? Embrace renewable energy options rather than relying fully on fossil fuels.

Solar panels can be pricey, but they are worth the investment if you want to save energy. Contrary to popular belief, you don't have to eliminate all electrical supplies from your electric company after transitioning. The sun can't provide power when it's dark, so you will still need a form of electricity for nighttime. Storms and colder weather can also limit the energy you receive from solar panels. Nonetheless, they are an excellent way to conserve energy and be kinder to the environment. Consider installing a solar generator alongside your panels to reduce the chances of a total outage.

Conclusion

In conclusion, the principles behind eco-friendly buildings and sustainable construction are simple and easy to apply. The aim is always to conserve natural resources and use resources efficiently while providing value to the local ecology and community and reducing the negative impacts on the environment associated with building development, operation, and eventual demolition.

In this guide, we have covered what it means for a building to be eco-friendly, what the features of green buildings are, the techniques and materials used to achieve the goals of sustainable constriction, and looked at some examples of the best eco-friendly buildings around the world.

We also looked at ways to make your building more eco-friendly, so you can get started right away! The causes for increasing utility bills can vary by household, but the overall inflation is relatively the same. There are actionable steps you can take to lower your utility bill and be more eco-friendly while doing so.

How to creating a Sustainable Future?

Existing building infrastructure generates over a quarter of annual global CO_2 emissions. In 2030, the entire building sector will only be allowed to emit a maximum of 72 million tons of CO_2 per year, which is cutting the emissions in half, according to 2019 values.

Companies face pressure and uncertainty from regulatory statutes, customers and internal stakeholders to make sustainable changes that meet these milestones. A sustainable lens is necessary to help reduce energy usage and choose more climate friendly materials along the building life cycle, from investment to design and from construction to operation.

To reach carbon neutrality, the industry will need to undergo digital transformation. Data transparency allows for the identification of energy efficiency potential. Companies must look towards solutions that address the entire life cycle of buildings to improve the total CO_2 footprint. These solutions will enable them to reach their long-term sustainability commitments.

Along the building life cycle, TÜV SÜD experts will contribute their expertise in advisory, testing, certification and inspection services to enable trust in new technologies and their journey into a sustainable future of their building portfolio.

At TÜV SÜD, we're prepared to support partners in new buildings and modernization projects.

10 Facts About Global Warming.

Global warming describes a change in the world's overall climate that results in rising temperatures over long-term periods of time and across the planet. Read on to learn 10 facts about global warming, including five effects of global warming.

1. Carbon Dioxide Leads Among Greenhouse Gases

Scientists agree that the main cause of global warming is greenhouse gas emissions, with carbon dioxide in first place ahead of other gas sources. One big contributor of greenhouse gas emissions is power stations, followed by industrial processes, transportation fuels and agricultural byproducts.

2. Animal Flatulence Creates Methane Gas

In addition to carbon dioxide, methane is a significant greenhouse gas contributing to climate change. Agricultural byproducts generate about 40 percent of methane emission, followed by 30 percent from fossil fuel retrieval, distribution and production, aCCOrding to the Emission Database for Global Atmospheric Research. Through normal digestion, animals such as cows, goats and sheep produce large methane amounts through their flatulence.

3. Deforestation Causes Global Warming

Because trees and plants help to regulate the climate by absorbing carbon dioxide and releasing oxygen, deforestation is a major cause of global warming. It reduces the number of trees available to release clean oxygen.

4. Ancient Evidence Shows Global Warming Exists

Ancient evidence such as tree rings, coral reefs, ocean sediments and sedimentary rock layers show that current global warming is happening. And it's occurring about 10 times faster than the average rate of warming during the Ice Age, aCCOrding to NASA.

5. Warmer Oceans Mean More Tsunamis

Ocean temperatures have risen nearly 20 degrees Fahrenheit since 1955, aCCOrding to the U.S. Environmental Protection Agency. Warmer ocean waters means tsunamis occur more frequently.

6. Global Warming Causes Glaciers to Melt

Not including Antarctica and Greenland's large ice sheets, the world has about 150,000 glaciers that cover about 200,000 square miles of Earth's surface. In the last 40 years, the glaciers have lost what's equal to a layer of ice 70 feet thick due to global warming-related causes, aCCOrding to a 2019 New York Times article. Air bubbles in glaciers provide a record of temperature and carbon dioxide stretching back 800,000 years, so scientists know the planet has experienced global warming before. But this "paleoclimate" evidence

Fig. - 9

also shows that the current warming is happening much more rapidly than in the past. The primary cause is the emission of greenhouse gases into the atmosphere, mostly carbon dioxide, which form a blanket that traps heat at the Earth's surface.

7. Fossil Fuels Contribute to Forest Fires

Fossil fuels possibly have created climate change consequences, such as London fog and forest fires in the United States. When fossil fuels are burned for human activities, it can warm Earth's atmosphere.

8. Global Warming Affects Hydroelectricity

Global warming affects hydroelectricity because it changes rivers and ecosystems. When their patterns change, they may not power dams the same ways they did before.

9. Polar Bears are Starving

Scientists warn that polar bears are at an increased risk of starving due to melting sea ice. As the ice melts, it becomes increasingly difficult for polar bears to hunt for seals.

10. Global Warming Causes Catastrophic Floods

Global warming contributes to rising sea levels and disasters such as catastrophic floods and flowing debris. Flooding can lead to economic disasters, too.

Global warming importance.

Global Warming is important since it helps determine future climate expectations. Through the use of latitude, one can determine the likelihood of snow and hail reaching the surface. You can also be able to identify the thermal energy from the sun that is accessible to a region. Global Warming is the scientific study of climates, which is defined as the mean weather conditions over a period of time. A branch of study within atmospheric sciences, it also takes into account the variables and averages of short-term and long-term weather conditions. Global warming is the unequivocal and continuing rise in the average temperature of Earth's climate system. Since 1971, 90% of the warming has occurred in the oceans. Despite the oceans' dominant role in energy storage, the term "global warming" is also used to refer to increases in average temperature of the air and sea at Earth's surface. Since the early 20th century, the global air and sea surface temperature has increased about 0.8 Â°C (1.4 Â°F), with about two-thirds of the increase occurring since 1980. Each of the last three decades has been successively warmer at the Earth's surface than any preceding decade since 1850. Global Warming uses from OMICS Group are an open access journal named as Journal of Climatology and Weather Forecasting which strives to release issues quarterly and is adamant to publish new findings related to the field of Global Warming. The mission of the Global Warming uses provides a forum for publishing new findings on Environmental principles and technology. Currently our primary research objective is to encourage and assist the development of better and faster measures of Environmental activity. In cases where we believe we can contribute directly, as opposed to through highlighting the work of others, we are producing our own measures of Global Warming.

Deforestation causes Global warming.

Because trees and plants help to regulate the climate by absorbing carbon dioxide and releasing oxygen, deforestation is a major cause of global warming. It reduces the number of trees available to release clean oxygen. Fossil fuels possibly have created climate change consequences, such as London fog and forest fires in the United States. When fossil fuels are burned for human activities, it can warm Earth's atmosphere.

Trees absorb carbon dioxide as they grow, acting as a "carbon sink". Cutting them down means more greenhouse gases entering the atmosphere, which speeds up the pace and severity of climate change. One of the important facts about global warming with respect to forests is that it still covers about 30% of the land, but some 50,000 square miles of forest are lost each year. That's equivalent to 48 football fields every minute. In the Amazon, for example, around 17% of the forest has been lost in the last 50 years.

Fig.-10

Global warming causes catastrophic Floods.

Global warming contributes to rising sea levels and disasters such as catastrophic floods and flowing debris. Flooding can lead to economic disasters, too.

The Paris Agreement.

Years in the making, the Paris Agreement, signed by 196 nations in 2015, aims to keep global temperature increase well below 2 degrees Celsius above pre-industrial levels, and if possible, below 1.5 degrees Celsius. This can only be achieved if countries stick to their commitments to significantly reduce greenhouse gas emissions. During his campaign, President Donald Trump promised to withdraw the US from the landmark agreement.

Fig.-11

Carbon Dioxide Emissions.

Air bubbles in glaciers provide a record of temperature and carbon dioxide stretching back 800,000 years, so scientists know the planet has experienced global warming before. But this "paleoclimate" evidence also shows that the current warming is happening much more rapidly than in the past. The primary cause is the emission of greenhouse gases into the atmosphere, mostly carbon dioxide, which form a blanket that traps heat at the Earth's surface. Human activities such as burning oil, coal and natural gas and deforestation have increased the amount of carbon dioxide by more than a third since the Industrial Revolution began.

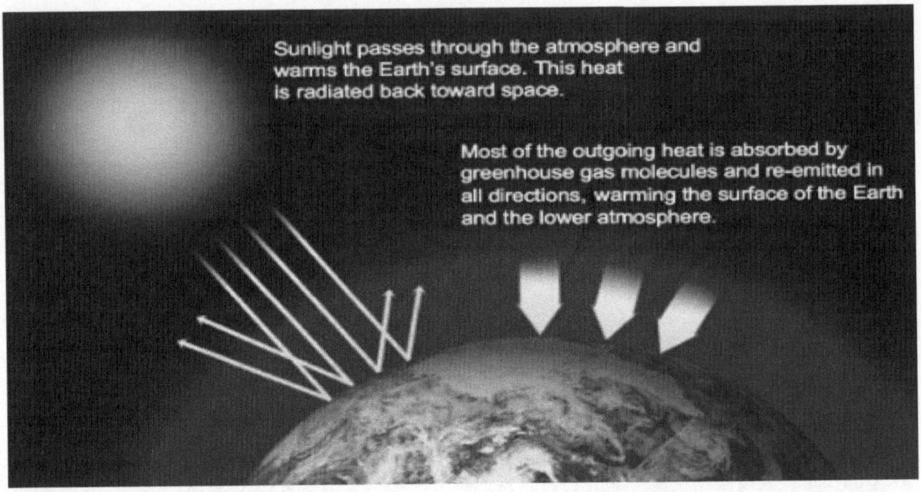

Fig.-12

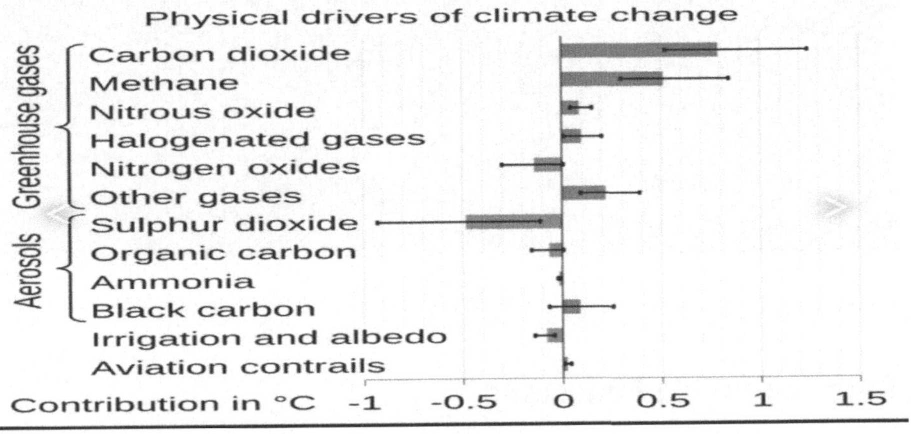

Fig.-13

Air bubbles in glaciers provide a record of temperature and carbon dioxide stretching back 800,000 years, so scientists know the planet has experienced global warming before. But this "paleoclimate" evidence also shows that the current warming is happening much more rapidly than in the past. The primary cause is the emission of greenhouse gases into the atmosphere, mostly carbon dioxide, which form a blanket that traps heat at the Earth's surface. Human activities such as burning oil, coal and natural gas and deforestation have increased the amount of carbon

The impact of global warming on humans and animals.

People are already suffering the consequences of climate change. Around 22.5 million people were displaced by climate or weather-related disasters between 2008 and 2015, according to the UN Refugee Agency (UNHCR). Climate change is also a factor in conflicts driving people from their homes which is an often ignored fact related to global warming. The UNHCR says that natural resources such as drinking water are likely to become more scarce and food security will become an even bigger concern in future because some crops and livestock won't survive in parts of the world if conditions become too hot and dry, or cold and wet. Climate change is also threatening wildlife: using satellite data from NASA, scientists estimate a possible 30% drop in the global population of polar bears over the next 35 years. That's because sea ice is their main habitat, and it is shrinking.

Fig.-14

The role of Deforestation in climate change and how can reducing Emissions from Deforestation and Degradation (REDD).

Deforestation refers to the purposeful clearing or thinning of trees and forests. When deforestation occurs, much of the carbon stored by trees is released back into the atmosphere as carbon dioxide, which contributes to climate change. In the last decade, the largest amounts of deforestation occurred across the humid tropics, mostly in Africa, followed by South America. The UN Food and Agriculture Organisation (FAO) estimates that around 420 million hectares of forest were lost between 1990 and 2020 (or 178 million hectares net, i.e. taking into account afforestation and the natural expansion of forests). The annual rate of deforestation has since slowed but was still 10 million hectares per year between 2015 and 2020. The most important driver of deforestation is the global demand for agricultural commodities: agribusinesses clear huge tracts of forest and use the land to plant high-value cash crops like palm oil and soya, and for cattle ranching. Land use change, principally deforestation, contributes 12–20% of global greenhouse gas emissions. Forest degradation (changes that negatively affect a forest's structure or function but that do not decrease its area), and the destruction of tropical peatlands, also contribute to these emissions. As a result of deforestation and degradation, some tropical forests now emit more carbon than they capture, turning them from a carbon 'sink' into a carbon source. For example, the south-eastern part of the Amazon Rainforest is now considered a net carbon source by scientists.

The scale of deforestation and its role in climate change:

Fig.-15

Deforestation refers to the purposeful clearing or thinning of trees and forests. When deforestation occurs, much of the carbon stored by trees is released back into the atmosphere as carbon dioxide, which contributes to climate change. In the last decade, the largest amounts of deforestation occurred across the humid tropics, mostly in Africa, followed by South America. The UN Food and Agriculture Organisation (FAO) estimates that around 420 million hectares of forest were lost between 1990 and 2020 (or 178 million hectares net, i.e. taking into account afforestation and the natural expansion of forests). The annual rate of deforestation has since slowed but was still 10 million hectares per year between 2015 and 2020. The most important driver of deforestation is the global demand for agricultural commodities: agribusinesses clear huge tracts of forest and use the land to plant high-value cash crops like palm oil and soya, and for cattle ranching.

Land use change, principally deforestation, contributes 12–20% of global greenhouse gas emissions. Forest degradation (changes that negatively affect a forest's structure or function but that do not decrease its area), and the destruction of tropical peatlands, also contribute to these emissions. As a result of deforestation and degradation, some tropical forests now emit more carbon than they capture, turning them from a carbon 'sink' into a carbon source. For example, the south-eastern part of the Amazon Rainforest is now considered a net carbon source by scientists.

Where does 'REDD+' come in?

Scientists have recognised the value of protecting forests in tackling climate change. In response, policymakers have developed a family of policies – collectively known as 'reducing emissions from deforestation and degradation' (REDD) – to provide a financial incentive to governments, agribusinesses and communities to maintain and possibly increase, rather than reduce, forest cover. The plus in 'REDD+' refers to "the role of conservation, sustainable management of forests and enhancement of forest carbon stocks in developing countries". Under REDD+, incentives for forest protection are offered to countries, communities and individual landowners in exchange for slowing deforestation, and carrying out activities that promote reforestation and sustainable forest management. Where local people are properly involved in the REDD+ process it may also help alleviate rural poverty. The principles of REDD+ were further reinforced in the Paris Agreement on climate change.

REDD+ policies operate through a variety of mechanisms, including those administered by the United Nations (UN-REDD) and the World Bank (the Forest Carbon Partnership Facility). REDD+ finance is also considered in the international climate change negotiations, remains a key component of international climate finance discussions, and is often channelled through the voluntary carbon markets and via activities implemented by for- and non-profit organisations.

How fair, effective and efficient is REDD+ :

While experts have demonstrated how REDD+ has the potential to reduce CO_2 emissions, it is not without its problems. For example, some question the fairness of a scheme that focuses on reducing emissions caused by some of the world's poorest people while emissions continue to rise in richer countries. Some developing countries may be wary of foreign interference in their land use policies. Researchers also highlight operational concerns – such as the difficulty in monitoring and measuring deforestation rates, or attributing changes in deforestation to REDD+ finance. Variations in local circumstances and institutional capacities mean that not all countries that have tropical forests possess the capabilities to address these challenges.

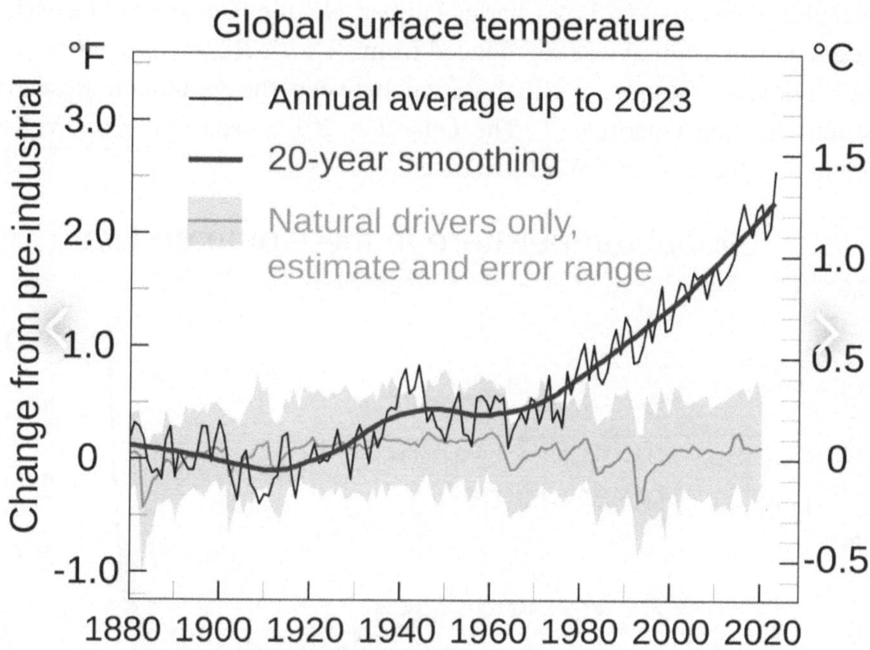

Fig.-16

How much REDD+ finance has been pledged :

Estimates of the global cost of REDD+ vary greatly, but at least US$15 billion would be needed annually to address tropical deforestation across the world. Current funding remains far off this mark: according to a 2020 review, between 2015 and 2019 an average of US$220 million a year of funding was approved. The Amazon Fund, with US$720 million of approved projects, remains the largest dedicated REDD+ fund. Without sufficient finance, it can be difficult to protect forests, as alternative land uses (such as for palm oil) can offer more immediate and guaranteed cash returns. Consequently, many experts have called for a scaling up of commitments and finance flows, although some have argued that even if large-scale REDD+ finance does materialise it may still struggle to compete with other land uses – especially as and when commodity prices rise. Whatever becomes of REDD+ in the future, experts agree it should focus first on areas that can most efficiently provide CO_2 reductions (such as tropical peat swamp forests) while also offering the potential for biodiversity conservation and poverty alleviation. This Explainer was updated in February 2023 by Charles Palmer, Natalie Pearson and Georgina Kyriacou. The original was reproduced from: *What's REDD and will it help tackle climate change?*, a collaboration between the Grantham Research Institute and the Guardian, © The Guardian 2012, used under a Creative Commons No Derivative Works licence.

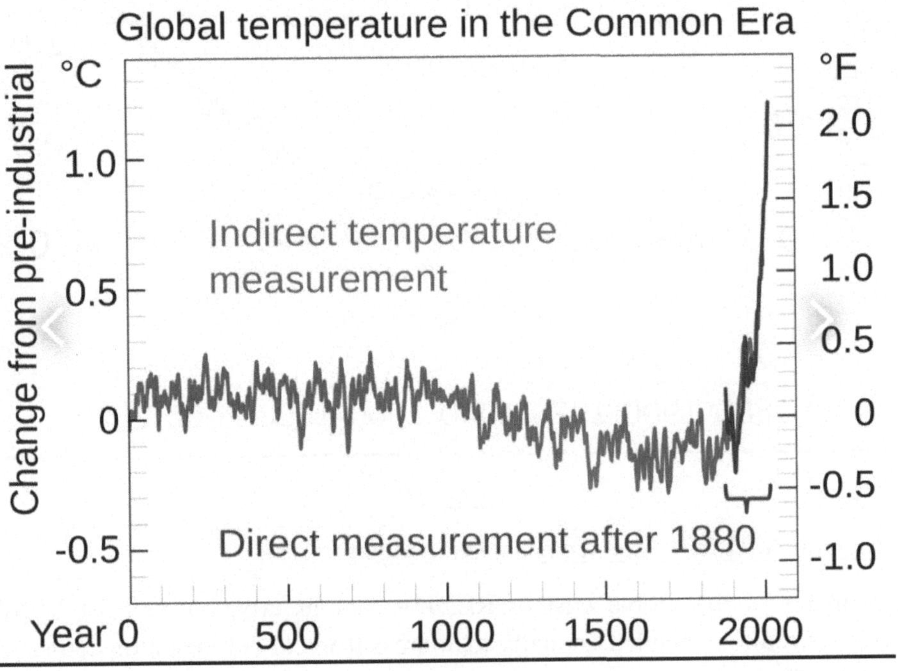

Fig.-17

How Deforestation Impacts Air Quality.

Breath is the foundation of any yoga or mindfulness practice. We harness the power of respiration to reduce stress and boost mental and physical performance. But what happens when the air we breathe is harmful to our health? The World Health Organization estimates that 91% of humans live where air quality levels are lower than recommended limits. And global deforestation is a major driver of rising air pollution.

Let's take a look at the problem with deforestation, how it contributes to air pollution, and what we can all do to breathe better today.

How Forests Affect Earth's Atmosphere.

Our atmosphere is a blend of gases — mostly nitrogen, a little oxygen, and micro-amounts of greenhouse gases like carbon dioxide. Greenhouse gases absorb the sun's rays and keep temperatures stable. Without them, we'd have sauna-like temps by day and frozen tundra temps by night. In these minuscule amounts, greenhouse gases are a good thing.

Forests help keep this process in balance. Through *sequestration*, trees capture and store carbon dioxide. Using light, water, and carbon dioxide, each tree produces energy in the form of sugar and releases oxygen into the air. During the day, trees also release water vapor along with oxygen, and atmospheric weather cycles move the clean air around the planet.

Forests are *carbon sinks*, which means they store more carbon dioxide than they release — twice as much, in fact. Each acre of forest sequesters two tons of carbon dioxide a year. That's equal to two cars' worth of yearly emissions.

When this process is in balance, global temperatures are regulated, and air pollution is minimized. But years of massive deforestation have tipped the balance toward unsustainable levels of greenhouse gas Emissions.

Deforestation and Air Pollution.

Deforestation involves the removal of large stands of trees for human use. Agriculture, raising livestock, mining, and drilling account for half of world forest loss. Other causes are urbanization, forest fires, and forestry practices like building roads.

Deforestation strips the planet of 18.7 million acres of forest a year, according to the World Wildlife Fund. That's equal to about 14 average-sized city parks every minute.

When vast areas of trees are stripped, stored carbon dioxide is released into the atmosphere, and those trees lose their carbon sequestration abilities. Worldwide, deforestation contributes to 15% of total greenhouse gas emissions.

Deforestation also increases the risk of forest fires. Clearcut areas of forest reduce humidity levels and lead to unstable ecosystems. Wildfire spreads faster and burns hotter, releasing more toxic pollutants into the air.

Every year, blazing forest fires release 8 billion tons of carbon dioxide. In 2015, Indonesia surpassed Canada and the United States in greenhouse gas emissions. The cause? Deforestation to make way for palm oil plantations.

Continued forest loss and increasing wildfires act in a feedback loop, emitting greenhouse gases and increasing air pollution around the world. Breath, the foundation of human life, is compromised.The good news is there are simple actions *anyone* can take to combat deforestation and improve the quality of the air we breathe.

How to take action to improve air quality and end Deforestation.

1. Support Indigenous rights to forest management by donating to organizations that work with Indigenous communities:

- Rainforest Action Network
- Amazon Conservation Team
- International Work Group for Indigenous Affairs

2. Support tree-planting efforts, and plant trees on your own property. One Tree Planted is an organization working in areas highly impacted by deforestation.
3. Reduce or eliminate meat consumption. Amazon rainforests are being cleared at rapid rates for cattle grazing. Consider a plant-based diet, and encourage others to do the same.
4. Avoid products made with palm oil, or look for the Roundtable on Sustainable Palm Oil (RSPO) certification. *Tip:* palm oil is most often found in processed foods like crackers, cookies, and other snacks. Always check the labels on these foods.
5. Choose bamboo paper products like Cloud Paper's paper towels and toilet paper — bamboo is a fast-growing sustainable alternative that absorbs massive amounts of carbon. In fact, it's the fastest growing

plant in the world and takes carbon out of the air quicker than almost any other plant.

These small actions can make a big difference. For better breathing and health, help us end deforestation today.

How Did Global Warming Start?

Most scientists agree that man-made climate change, or global warming, began in earnest during the industrial revolution, which occurred between the 18th and 19th centuries. This occurred, they say, as a result of a large number of greenhouse gases released into the atmosphere. However, in 2008, researchers from the University of Wisconsin-Madison concluded that global warming began as many as 8,000 years earlier with the advent of agriculture and deforestation.

Greenhouse gases released into the atmosphere during and since the industrial revolution remain trapped, where they continue to accumulate. They also serve to retain heat, which has had a series of effects on the planet's climate.

Early in the 20th century, most scientists and industrial leaders believed it was possible to alter a local climate but not the global climate, based on local activities. Although some scientists, including Svante Arrhenius, a Swedish chemist, saw the possibility for global climate change, they were dismissed as being alarmists.

Beginning in the late 1950s, scientists observed a steady increase in carbon dioxide in the atmosphere based on precise measurements. By the 1980s, the global temperature had begun to increase to such a degree that the media heard about it and began to spread the news.

Exploring the Sun's Role in Climate Change and Global Warming.

The sun is the ultimate source of energy for life on Earth. It provides light, heat, and sustains all living organisms. However, recent research has shown that the sun may also play a role in climate change and global warming. In this article, we will explore the connection between the sun's activity and its impact on our planet's climate.

The Sun's Energy Output

The sun emits vast amounts of energy in the form of electromagnetic radiation. This energy is generated through nuclear fusion reactions taking place within its core. The amount of energy released by the sun varies over time due to natural fluctuations in its activity. Scientists measure solar activity by monitoring several key indicators, including sunspot numbers, solar irradiance, and solar flares. Sunspots are dark patches on the surface of the sun that indicate areas of intense magnetic activity. Solar irradiance refers to the amount of solar energy reaching Earth's atmosphere.

Solar Cycles and Climate Patterns

Solar activity follows an 11-year cycle known as the solar cycle or sunspot cycle. During periods of high solar activity, more sunspots are observed, indicating increased magnetic turbulence on the surface of the sun. Conversely, during periods of low solar activity, fewer sunspots are seen.

Several studies have suggested a correlation between solar cycles and climate patterns on Earth. When solar activity is high, scientists have observed a slight increase in global temperatures. This can be attributed to higher levels of solar irradiance reaching our planet.

Solar Forcing and Its Impact on Climate

Solar forcing refers to changes in Earth's climate caused by variations in solar radiation received at its surface. While greenhouse gases emitted by human activities are considered major contributors to global warming, scientists believe that changes in solar forcing can also influence our climate system. During periods of increased solar activity, more radiation from the sun reaches Earth's atmosphere. This additional energy can lead to a slight warming effect. However, the impact of solar forcing on climate is relatively small compared to the long-term effects of greenhouse gas emissions.

Future Research and Implications

Understanding the sun's role in climate change and global warming is crucial for predicting future climate patterns accurately. Ongoing research aims to improve our understanding of solar activity and its impact on

Earth's climate system. While the sun does contribute to climate variations, it is essential to note that human activities, such as burning fossil fuels and deforestation, are the primary drivers of global warming. The increase in greenhouse gas emissions has far surpassed any natural fluctuations caused by the sun. In conclusion, the sun's activity does play a role in climate change and global warming to some extent. Solar cycles and variations in solar forcing can lead to minor fluctuations in Earth's climate patterns. However, human-induced factors still dominate the long-term trends observed in global temperatures. It is vital for us to reduce greenhouse gas emissions and adopt sustainable practices to mitigate the effects of anthropogenic climate change.

What Are the Positive Effects of Global Warming?

Positive effects of global warming include fewer deaths in winter due to cold weather, lower costs of energy for heating homes and other buildings, and better agricultural production. Additionally, warmer temperatures facilitate sea transportation in the northern parts of the Earth.

In most parts of the world, deaths from cold in winter exceed deaths from heat in summer. Winter temperatures especially impact the poor, who cannot afford to sufficiently heat their homes. A rise of even a few degrees of temperature annually decreases fuel expenditures, saves lives and reduces medical costs. Additionally, many people prefer warmer climates, as evidenced by the tendency of retirees to move to southern locations. Global warming also increases levels of carbon dioxide in the atmosphere. Plants thrive in areas with high carbon dioxide concentrations, and crops that absorb more carbon dioxide may grow larger and have increased yields. Warmer climates also open up land for agricultural use in areas that were previously too cold for farming.

In the north, global warming enables commercial fishing at higher latitudes and opens up previously closed areas for shipping. For instance, the fabled Northwest Passage in northern Canada can become a reality, enabling ocean-going traffic to move from the Pacific Ocean to the Atlantic Ocean and back by a northern route.

UNOPS Announces Climate Commitment.

At an accelerated rate, climate change is impacting communities in every country, across every continent. From rising sea levels and extreme weather events, to energy, food and water insecurity – the consequences are clear, and they are tragic.

Developing countries and vulnerable populations who are on the front lines of climate change need support. Yet global efforts to combat climate change and achieve sustainable development are falling short.

UNOPS is committed to providing practical solutions to help countries combat climate change, through our expertise in infrastructure, procurement and project management services.

We believe infrastructure is at the heart of building a sustainable, resilient and inclusive future, and is key to ensuring access to clean and affordable energy to all.

While sustainable supply chains can help protect natural resources and limit environmental impacts, the sheer scale and volume of public procurement spending makes it a powerful catalyst for advancing climate action Across the globe, we are committed to supporting mitigation and adaptation efforts, and as efforts continue to scale up development and climate finance, we are committed to supporting countries to implement and operationalize the Sustainable Development Goals and the Paris Agreement.

There is a significant implementation gap that we must bridge to respond to the immense needs brought on by the climate emergency. The time to react and adapt is now.

What is Global Warming?

Global warming is the phenomenon of gradual increase in the average temperature of earth. It is caused by the release of greenhouse gases like carbon dioxide, methane, CFCs etc. into the atmosphere.

Global warming is the phenomenon of a gradual increase in the temperature near the earth's surface. This phenomenon has been observed over the past one or two centuries. This change has disturbed the climatic pattern of the earth. However, the concept of global warming is quite controversial but the scientists have provided relevant data in support of the fact that the temperature of the earth is rising constantly.

There are several causes of global warming, which have a negative effect on humans, plants and animals. These causes may be natural or might be the outcome of human activities. In order to curb the issues, it is very important to understand the negative impacts of global warming.

The gradual increase in temperature due to the greenhouse effect caused by pollutants, CFCs and carbon dioxide is called global warming. This phenomenon has disturbed the climatic pattern of the earth.

The burning of fossil fuels, emissions from automobiles and chlorofluorocarbons add to the greenhouse gases in the atmosphere. This has led to an increase in the earth's temperature causing environmental changes. This increase in temperature across the globe is known as global warming.

Causes of Global Warming?

Following are the major causes of global warming:

Man-made Causes of Global Warming

Deforestation

Plants are the main source of oxygen. They take in carbon dioxide and release oxygen thereby maintaining environmental balance. Forests are being depleted for many domestic and commercial purposes. This has led to an environmental imbalance, thereby giving rise to global warming.

Use of Vehicles

The use of vehicles, even for a very short distance results in various gaseous emissions. Vehicles burn fossil fuels which emit a large amount of carbon dioxide and other toxins into the atmosphere resulting in a temperature increase.

Chlorofluorocarbon

With the excessive use of air conditioners and refrigerators, humans have been adding CFCs into the environment which affects the atmospheric ozone layer. The ozone layer protects the earth surface from the harmful ultraviolet rays emitted by the sun. The CFCs have led to ozone layer depletion making way for the ultraviolet rays, thereby increasing the temperature of the earth.

Industrial Development

With the advent of industrialization, the temperature of the earth has been increasing rapidly. The harmful emissions from the factories add to the increasing temperature of the earth.

In 2013, the Intergovernmental Panel for Climate Change reported that the increase in the global temperature between 1880 and 2012 has been 0.9 degrees Celsius. The increase is 1.1 degrees Celsius when compared to the pre-industrial mean temperature.

Agriculture

Various farming activities produce carbon dioxide and methane gas. These add to the greenhouse gases in the atmosphere and increase the temperature of the earth.

Overpopulation

An increase in population means more people breathing. This leads to an increase in the level of carbon dioxide, the primary gas causing global warming, in the atmosphere.

Natural Causes of Global Warming:

Volcanoes

Volcanoes are one of the largest natural contributors to global warming. The ash and smoke emitted during volcanic eruptions goes out into the atmosphere and affects the climate.

Water Vapour

Water vapour is a kind of greenhouse gas. Due to the increase in the earth's temperature, more water gets evaporated from the water bodies and stays in the atmosphere adding to global warming.

Melting Permafrost

Permafrost is frozen soil that has environmental gases trapped in it for several years and is present below Earth's surface. It is present in glaciers.

As the permafrost melts, it releases the gases back into the atmosphere, increasing Earth's temperature.

Forest Blazes

Forest blazes or forest fires emit a large amount of carbon-containing smoke. These gases are released into the atmosphere and increase the earth's temperature resulting in global warming.

Effects of Global Warming.

Following are the major effects of global warming:

Rise in Temperature

Global warming has led to an incredible increase in earth's temperature. Since 1880, the earth's temperature has increased by ~1 degrees. This has resulted in an increase in the melting of glaciers, which have led to an increase in the sea level. This could have devastating effects on coastal regions.

Threats to the Ecosystem

Global warming has affected the coral reefs that can lead to the loss of plant and animal lives. Increase in global temperatures has made the fragility of coral reefs even worse.

Climate Change

Global warming has led to a change in climatic conditions. There are droughts at some places and floods at some. This climatic imbalance is the result of global warming.

Spread of Diseases

Global warming leads to a change in the patterns of heat and humidity. This has led to the movement of mosquitoes that carry and spread diseases.

High Mortality Rates

Due to an increase in floods, tsunamis and other natural calamities, the average death toll usually increases. Also, such events can bring about the spread of diseases that can hamper human life.

Loss of Natural Habitat

A global shift in the climate leads to the loss of habitats of several plants and animals. In this case, the animals need to migrate from their natural habitat and many of them even become extinct. This is yet another major impact of global warming on biodiversity.

What do CFCs stand for? What is the role of CFC in global warming?

CFCs stand for chlorofluorocarbons. Ozone layer is responsible for protecting the surface of the earth from the sun's harmful radiations. CFCs destroy the ozone layer of the atmosphere. This makes the way for the ultraviolet rays to reach the earth, thus, increasing the temperature which leads to global warming.

How does global warming affect climate change?

The change in climatic conditions is a result of global warming. The burning of fossil fuels, cutting down of trees etc. causes the temperature of the earth to increase. High temperature changes the weather patterns, causing the dry areas to get drier and wet areas to get wetter. Thus, increasing the frequency of disasters like floods, droughts etc.

How can we control global warming?

The release of carbon dioxide and other greenhouse gases into the atmosphere is the major cause of global warming. It can be reduced by setting a high price of carbon, increasing the biofuels production from organic waste, use of renewable energy like solar and wind power, safeguarding forests and improving energy efficiency and vehicle fuel economy.

To know more about global warming, its definition, causes and effects, keep visiting the BYJU'S website or download BYJU'S app for further reference.

Greenhouse Effect Definition.

"Greenhouse effect is the process by which radiations from the sun are absorbed by the greenhouse gases and not reflected back into space. This insulates the surface of the earth and prevents it from freezing."

What is the Greenhouse Effect?

A greenhouse is a house made of glass that can be used to grow plants. The sun's radiations warm the plants and the air inside the greenhouse. The heat trapped inside can't escape out and warms the greenhouse which is essential for the growth of the plants. Same is the case in the earth's atmosphere.

During the day the sun heats up the earth's atmosphere. At night, when the earth cools down the heat is radiated back into the atmosphere. During this process, the heat is absorbed by the greenhouse gases in the earth's atmosphere. This is what makes the surface of the earth warmer, that makes the survival of living beings on earth possible.

However, due to the increased levels of greenhouse gases, the temperature of the earth has increased considerably. This has led to several drastic effects.

Let us have a look at the greenhouse gases and understand the causes and consequences of greenhouse effects with the help of a diagram shown in LVII.

Greenhouse Gases and Greenhouse Effect Diagram.

"Greenhouse gases are the gases that absorb the infrared radiations and create a greenhouse effect. For eg., carbon dioxide and chlorofluorocarbons."

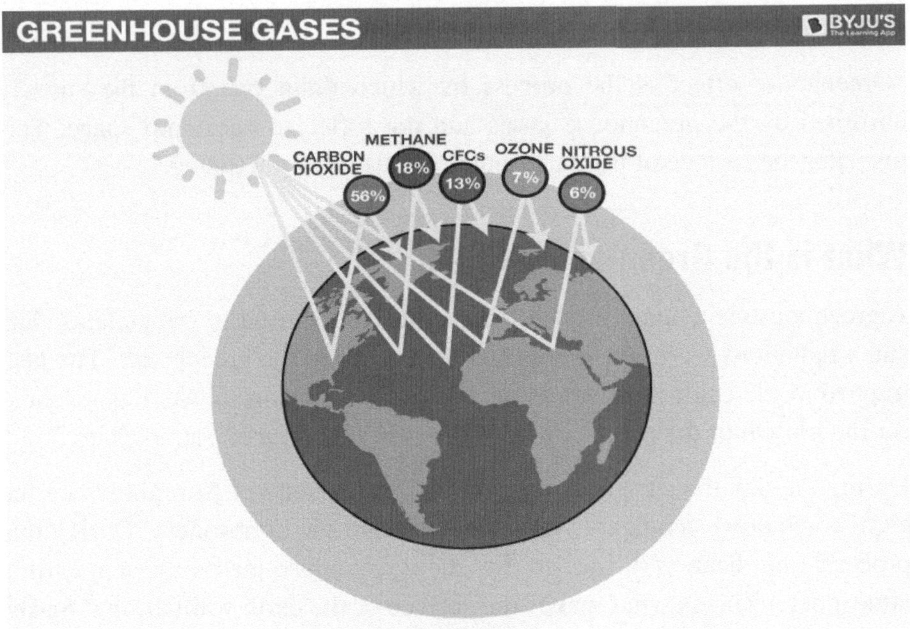

Fig.-18

The Diagram shows Greenhouse Gases such as carbon dioxide are the primary cause for the Greenhouse Effect

The major contributors to the greenhouse gases are factories, automobiles, deforestation, etc. The increased number of factories and automobiles increases the amount of these gases in the atmosphere. The greenhouse gases never let the radiations escape from the earth and increase the surface temperature of the earth. This then leads to global warming.

Causes of Greenhouse Effect.

The major causes of the greenhouse effect are:

Burning of Fossil Fuels

Fossil fuels are an important part of our lives. They are widely used in transportation and to produce electricity. Burning of fossil fuels releases carbon dioxide. With the increase in population, the utilization of fossil fuels has increased. This has led to an increase in the release of greenhouse gases in the atmosphere.

Deforestation

Plants and trees take in carbon dioxide and release oxygen. Due to the cutting of trees, there is a considerable increase in the greenhouse gases which increases the earth's temperature.

Farming

Nitrous oxide used in fertilizers is one of the contributors to the greenhouse effect in the atmosphere.

Industrial Waste and Landfills

The industries and factories produce harmful gases which are released in the atmosphere.

Landfills also release carbon dioxide and methane that adds to the greenhouse gases.

Effects of Greenhouse Effect.

The main effects of increased greenhouse gases are:

Global Warming

It is the phenomenon of a gradual increase in the average temperature of the Earth's atmosphere. The main cause for this environmental issue is the increased volumes of greenhouse gases such as carbon dioxide and methane released by the burning of fossil fuels, emissions from the vehicles, industries and other human activities.

Depletion of Ozone Layer

Ozone Layer protects the earth from harmful ultraviolet rays from the sun. It is found in the upper regions of the stratosphere. The depletion of the ozone layer results in the entry of the harmful UV rays to the earth's surface that might lead to skin cancer and can also change the climate drastically.

The major cause of this phenomenon is the accumulation of natural greenhouse gases including chlorofluorocarbons, carbon dioxide, methane, etc.

Smog and Air Pollution

Smog is formed by the combination of smoke and fog. It can be caused both by natural means and man-made activities.

In general, smog is generally formed by the accumulation of more greenhouse gases including nitrogen and sulfur oxides. The major contributors to the formation of smog are automobile and industrial emissions, agricultural fires, natural forest fires and the reaction of these chemicals among themselves.

Acidification of Water Bodies

Increase in the total amount of greenhouse gases in the air has turned most of the world's water bodies acidic. The greenhouse gases mix with the rainwater and fall as acid rain. This leads to the acidification of water bodies.

Also, the rainwater carries the contaminants along with it and falls into the river, streams and lakes thereby causing their acidification.

Runaway Greenhouse Effect

This phenomenon occurs when the planet absorbs more radiation than it can radiate back. Thus, the heat lost from the earth's surface is less and the temperature of the planet keeps rising. Scientists believe that this phenomenon took place on the surface of Venus billions of years ago.

This phenomenon is believed to have occurred in the following manner:

- A runaway greenhouse effect arises when the temperature of a planet rises to a level of the boiling point of water. As a result, all the water from the oceans converts into water vapour, which traps more heat coming from the sun and further increases the planet's temperature. This eventually accelerates the greenhouse effect. This is also called the "positive feedback loop".
- There is another scenario giving way to the runaway greenhouse effect. Suppose the temperature rise due to the above causes reaches such a high level that the chemical reactions begin to occur. These chemical reactions drive carbon dioxide from the rocks into the atmosphere. This would heat the surface of the planet which would further accelerate the transfer of carbon dioxide from the rocks to

the atmosphere, giving rise to the runaway greenhouse effect. In simple words, increasing the greenhouse effect gives rise to a runaway greenhouse effect which would increase the temperature of the earth to such an extent that no life will exist in the near future.

List gases which are responsible for the greenhouse effect.

The major greenhouse gases are:
1) Carbon dioxide
2) Methane
3) Water
4) Nitrous oxide
5) Ozone
6) Chlorofluorocarbons (CFCs)

What would have happened if the greenhouse gases were totally missing in the earth's atmosphere?

Earth's most abundant greenhouse gases are carbon dioxide, methane, ozone, CFCs, and water vapour. These gases are present near the Earth's surface. They absorb solar energy that is radiated back from the surface of the Earth. The absorption of radiation results in the heating up of the atmosphere. Hence, greenhouse gases are essential for maintaining the temperature of Earth for the sustenance of life.
In the absence of greenhouse gases, the average temperature of Earth will decrease drastically, making it uninhabitable. As a result, life on Earth would be impossible.

Development history of Green Buildings.

1. In the 1960s, American architect Paul Soleri proposed a new concept of ecological architecture.
2. In 1969, American architect Ian McHarg wrote the book "Design Integrates Nature", which marked the official birth of ecological architecture.

3. In the 1970s, the energy crisis caused various building energy-saving technologies such as solar energy, geothermal energy and wind energy to emerge, and energy-saving buildings became the forerunner of building development.
4. In 1980, the World Conservation Organization put forward the slogan "sustainable development" for the first time. At the same time, the energy-saving building system was gradually improved, and it was widely used in developed countries such as Germany, Britain, France and Canada.
5. In 1987, the United Nations Environment Program published the "Our Common Future" report, which established the idea of sustainable development.
6. In 1990, the world's first green building standard was released in the UK.
7. In 1992, because the "United Nations Conference on Environment and Development" promoted the idea of sustainable development, green buildings gradually became the direction of development.
8. In 1993, the United States created the Green Building Association.
9. In 1996, Hong Kong introduced green building standards.
10. In 1999, Taiwan introduced green building standards.
11. In 2000, Canada introduced green building standards.
12. In 2005, Singapore initiated the "BCA Green Building Mark"
13. In 2015, according to the Berkeley National Laboratory, China implemented the "Green Building Evaluation Standards.
14. In 2021, the first, both low-cost and sustainable 3D printed house made out of a clay-mixture was completed.

Green building by country.

1. Green building in Australia
2. Green building in Bangladesh
3. Green building in Germany

4. Green building in Israel
5. Green building in South Africa
6. Green building in the United Kingdom
7. Green building in India
8. Green building in the United States
9. The Model home 2020 project: Denmark, Austria, Germany, France, UK

CHAPTER - 2

ECO-FRIENDLY BUILDING CONSTRUCTION BY PRECAST CONCRETE TECHNOLOGY METHODS:

Abstract.

The precast concrete technology system is very versatile and is equally efficient and suitable for all types of construction, be it is a high rise or low-rise buildings, villas, commercial buildings, parking lots etc. Most of the construction activities in India take place by conventional cast in situ method of construction. But still there is a huge demand for housing in India. So, the construction activity has to take place in a much faster way.by producing precast concrete in a controlled environment (typically referred to as a Precast plant), the precast concrete is afforded the opportunity to properly cure and be closely monitored by plant employees. This cannot be achieved by conventional method of construction. It can be done possible with precast concrete of construction. Moreover, there are more advantages of precast concrete when compared with conventional one. So various literatures are studied and a review of those all has been given in this paper. Also, the advantages and disadvantages of precast construction are also discussed here. Precast elements can be quickly erected on the job site and precast thermal mass can also save energy and increase comfort. Major reasons and advantages to adopt this technology are structural stability, speedy and quality construction, flexibility, wastage control and less manpower requirement. Other than this, the strength of this technology is, low maintenance, seismic resistance and universal application. Precast concrete system which is mass produced either in factory or at site factory is widely used as building components. Today, the world agrees that precast construction makes significant impact on both time and budget. This was done by analyzing the scheduled activities and budget of the project and implementing it in real case studies. The components are usually designed according to the specified standard shapes and dimensions. The components will then be transported to the construction site to be placed according to the building design requirements. This Technology Promises us a world of creative possibilities in application and design by offering an abundance of choice for the Architects and Engineers to create building unique in character and of exceptionally high Quality Components like roof slabs, Beams, columns, wall panels etc. offers flexibility in shape and size with a variety of surface finishes and colours.

(A) Mass- Precast concrete products can act as effective barriers to vehicular traffic due to their size and weight. In the current world climate, precast concrete products such as planters, bollards and highway barriers increasingly are being used to provide protection for a wide variety of venues.

(B) Chemical Resistance- Precast is resistant to most substances. However no material is completely immune to attack form aggressive chemical agents. Thus it is wise to choose the material with the longest expected service life Precast concrete products can be designed to withstand anticipated corrosive agents.

(C) Quality Control- Because precast concrete products typically are produced in a controlled environment they exhibit high quality and uniformity. Variables affecting quality typically found on a jobsite-temperature, humidity material quality craftsmanship-are nearly eliminated in a plant environment.

(D) UV Sensitivity- Unlike some other some other materials, precast concrete does not degrade from exposure to sunlight. This is extremely beneficial for above-ground applications.

(E) Environmentally Friendly - After water, concrete is the most frequently used material on earth, it is nontoxic, environmentally safe and composed of natural materials. Buried throughout the world, precast concrete products help convey water contributing to poor water quality.

(F) Weather Resistance- Precast concrete is well-suited for exposures to all types of weather conditions. Reduced Weather Dependency - Precast increases efficiency because weather will not delay production. In additional weather conditions at the jobsite do not significantly affect the schedule. This is because it requires less time to install precast compared with other construction methods, such as cast-in-place concrete. Precast concrete can be easily installed on demand and immediately backfilled-there is no need to wait for it to cure.

(G) Water Tightness - Standard watertight sealants are specially formulated to adhere to precast concrete, Making watertight multiple-seam precast concrete structures possible.

(H) Ease of Installation- Since precast products are designed and manufactured for simple connection, many components can be installed in a short time.

(I) Modularity - Because of the modular nature of many precast concrete products, structures of systems of nearly any size can be installed in a short time.

(J) Efficiency- Precast concrete products arrive at the jobsite ready to install. There is no need to order raw materials such as reinforcing steel and concrete, and there is no need to spend time setting up forms, placing concrete or waiting for the concrete to cure.

(K) Aesthetics - Precast concrete products are both functional and decorative. They can be shaped and molded into an endless array of size and configurations Precast concrete can also be produced in virtually any color and a wide variety of finishes (acid-etched. stand blasted, smooth-as-cast. exposed-aggregate) to achieve t he desired appearance f or building and site application.

(L) Low Maintenance- Precast concrete requires or on maintenance, which makes it an ideal choice for nearly any design solution. Fire Resistance- Precast concrete is noncombustible, also, concrete does not its structural capacity nearly as quickly as steel.

(M)Fire Resistance- Precast concrete is noncombustible, also, concrete does not its structural capacity nearly as quickly as steel. Other materials besides concrete and steel are flammable and / or do not perform well in elevated temperature.

Introduction.

Precast concrete is widely used in the construction industry. Handling, transporting and erecting precast concrete elements are high-risk activities that have resulted in deaths and serious injuries to workers. Every person working with precast concrete has health and safety duties. Everyone should clearly understand their roles and responsibilities when working with precast concrete. Important decisions at design stages determine the success of the project in terms of speed and quality of construction, environmental impact and aesthetic appeal of the structure, as well as, cost and economic viability. This paper discusses some of the challenges faced in the viaduct structures and how they were overcome in building Constructions in India by the employment of Precast Concrete Construction Technologies. Nowadays the precast concrete has been used extensively in many residential and commercial construction projects. It is because of the property of the precast concrete. It has higher durability, thermal properties and very easy to handle and so on. Also, the quality of the precast concrete is higher as it is manufactured under great factory-controlled environment. But there is lack of awareness and knowledge regarding the precast concrete in India. This has to be changed and more study has to be done regarding precast. The connection details of the precast members also should be studied well. In this journal various literatures are studied and the results of various studies have been given here. Precast Concrete is a very sustainable building material with regard to the energy consumption during its production and its characteristics during its useful life. Concrete is a natural building material and 100% recyclable. Concrete consists of water, sand, gravel and cement. Water, sand and gravel naturally exist in our environment.

Wall Panels

Parapet Beams

Spandrel

Solid Slab Panels

Pod Elements

Staircase

Fig.-18

Advantages and Dis-Advantages of Precast Concrete Technology

Advantages of Precast Concrete:

There are many precast concrete advantages. They are discussed below.

A. Saves Construction Time: Precast Concrete construction saves time, and the risk of project delay is also less. The precast concrete casting can be carried on simultaneously with other works on site such as earthwork, survey, etc., and thus saves time. It is a major advantage of precast concrete.

B. Quality Assurance: The key factors which regulate the quality of construction such as curing, temperature, mix design, formwork, etc. can be monitored for pre cast concrete. So, improved quality construction can be performed.

C. Usage of Pre-stressed Concrete: By using pre-stressed precast, structural materials of high strength and load-bearing capacity can be achieved, which can result in greater clear span, reduced size of the cross-section of structural members, etc.

D. Cost-effective: The simplified construction process reduces time, increases productivity, quality and safety and thus the cost is reduced.

E. Durability: Precast Concrete structure has a longer service time period and minimal maintenance. The high-density Precast Concrete is more durable against acid attack, corrosion, and impact, reduces surface voids, and resists the accumulation of dust.

F. Aesthetics: As the structures are made of prefabricated concrete in a controlled factory environment, several combinations of colors and textures can be used. A wide range of shapes and sizes are available to choose from with smooth finishing and thus the aesthetical value of products is increased.

G. Safe Construction Platform: No raw materials have to be stocked on the site for precast concrete construction. It reduces the requirement of traditional formworks and props, wastage, workers, etc., and thus provides a safe working platform.

Dis-Advantages of Precast Concrete:

There are some disadvantages to precast concrete. They are discussed below. High Initial Investment For installing a precast concrete plant, heavy and sophisticated machines are necessary which requires a high initial investment. A large scale of precast construction projects must be available to ensure sufficient profit.

A. Transportation Issue: The construction site can be at a distant location from the precast concrete plant. in that case, the precast members must be carried to the site using trailers. in many cases, the reduced costs of pre cast concrete are compensated by the transportation cost.
B. Handling Difficulties: Proper care and precaution have to be taken for handling precast concrete. Usually, precast members are heavy and large which makes them difficult to handle without damage. Generally, portable or tower cranes are used to handle precast members.
C. Modification: Limitation in the case of precast structures, it is difficult to modify the structure. For example, if structural precast concrete walls are to be dismantled for modification it will impact the overall stability of the structure.
D. Sensitive Connection Works: Assembling the precast members is one of the key points for ensuring strong structural behavior. Connections between several structural members must be supervised and done properly to ensure the intended behavior of the connection such as simple, semi-rigid, or rigid connections. Besides this, faulty connections in precast concrete may lead to water leakage and failed sound insulation.

Precast Concrete Construction System in Brief.

Precast concrete construction system has its own characteristics which influence the layout, span length, construction depth, and stability system to a great extent. In precast concrete construction, majority of structural members are manufactured in manufacturing plants away from the construction site. After that, they are delivered to the project site to be erected. Various connections have been developed to connect different types of structural elements like beam to column connection and panel to panel connection. In this manner, the imposed loads are transferred from superstructure to the foundation. Designers should consider the possibilities, restrictions, detailing, manufacturer, transport, erection and serviceability stages before finalizing the design of precast concrete structure. Large panel systems are useful for the construction of apartments and hotels. It consists of large walls and floor concrete panels connected in the vertical and horizontal directions. Both horizontal and vertical panels withstand gravity loads. There are three arrangements of large panel system based on wall layouts which include cross-wall system, longitudinal-wall system, two-way system.

Types of Precast Elements.

Two main types of precast concrete elements, namely precast reinforced concrete elements and precast pre-stressed concrete elements are used as per the details given below: Precast concrete elements – Concrete components of a building prefabricated in precast yard or site and shall be installed in the building during construction.

i. Precast reinforced concrete elements

   These shall consist of reinforcement bars and welded wire meshes within the elements to provide the tensile strength and resistance against cracks such as façade walls, beams, columns, slabs, refuse chutes, staircases and parapet walls.

ii. Precast pre-stressed concrete elements

These shall consist of pre-stressing tendons within the elements to provide a predetermined force needed to resist external loadings and cracks such as hollow core slabs, beams and panels.

Types of Precast Moulds.

1. Mould Types - Precast Wall:

Structural load bearing walls shall be designed as per codal provisions of IS 456:2000 and IS 13920:1993 as applicable. Internal non load bearing walls should be designed as plain concrete walls with nominal reinforcement for handling and erection stresses. Such walls may also be built using alternate partition wall systems. For concrete walls, minimum concrete grade should be M10.

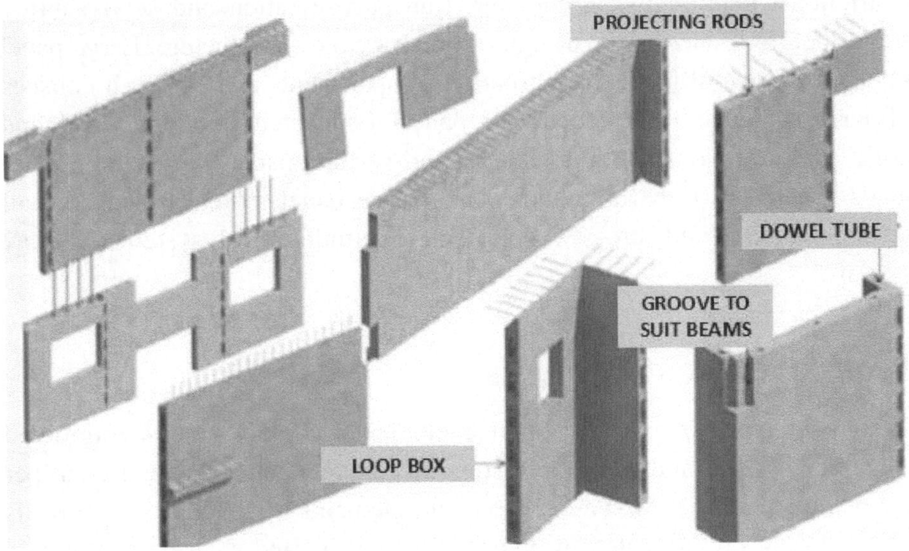

Fig.19- Mould types - Precast Wall

2. Mould Types - Precast Pods:

Pod technology is a type of offsite modular construction that has been there in the market since 1950s and has consistently gained popularity over time. Offsite construction of pods completed in a factory capitalizes on the ability to move product in a controlled manufacturing condition, on tight inventory control and project schedules. It is inherently waste conscious and can have minimum site impact if delivered carefully and strategically with respect to site constraints. These offsite solutions negate the need for snagging the finished bathroom areas – reducing the time and cost associated with what is widely known to be one of the most challenging components of a building to deliver. These self-supporting structures are factory made and assembled. This includes all mechanical, electrical and plumbing fixtures with termination of services at pre-agreed points.

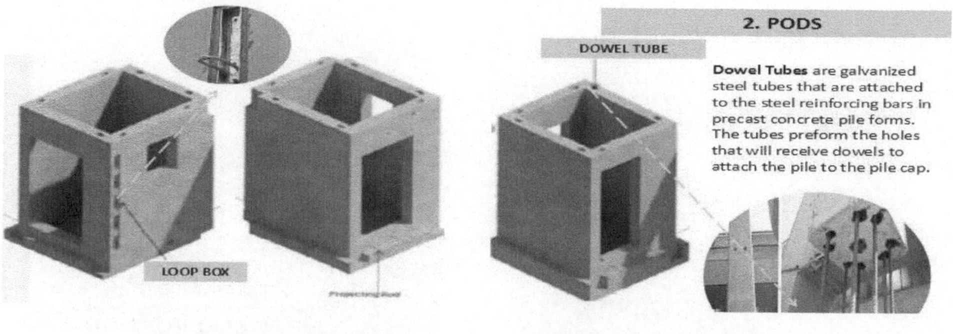

Fig – 20 - Mould types - Precast Pods

3. Mould types - Precast Spandrels Beam

Spandrels are horizontal precast pieces that may be structural with pockets, corbels or ledges to support double tees and hollow core plank, or non-structural elements in total precast buildings.

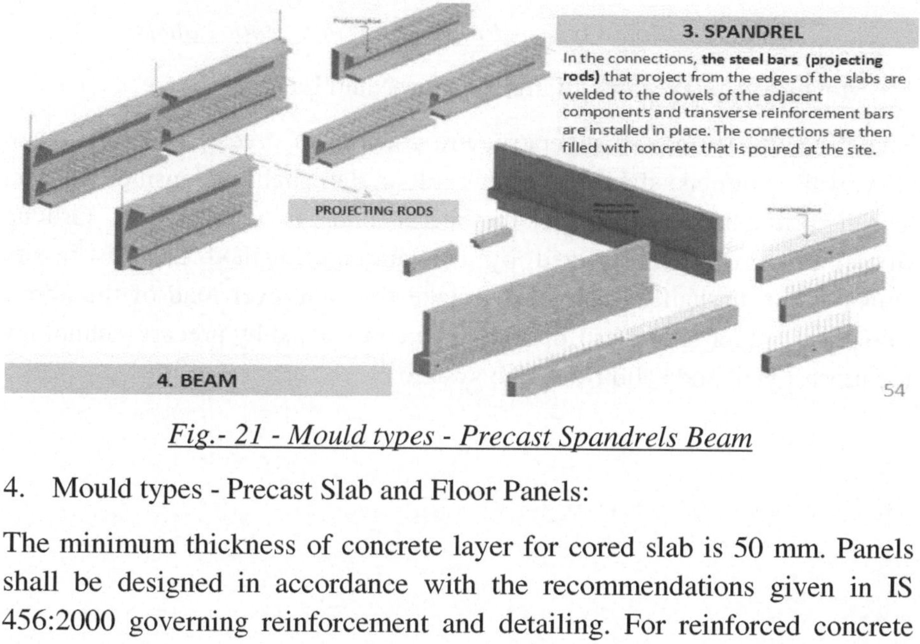

Fig.- 21 - Mould types - Precast Spandrels Beam

4. Mould types - Precast Slab and Floor Panels:

The minimum thickness of concrete layer for cored slab is 50 mm. Panels shall be designed in accordance with the recommendations given in IS 456:2000 governing reinforcement and detailing. For reinforced concrete floors, concrete of minimum grade M 30 shall be used. Pre-stressed concrete floor units shall satisfy the strength requirements followed in usual design practice, namely, a minimum of M 35 for post-tensioned and pre-tensioned works. The thickness of the floor panel shall be such that the serviceability requirements are satisfied.

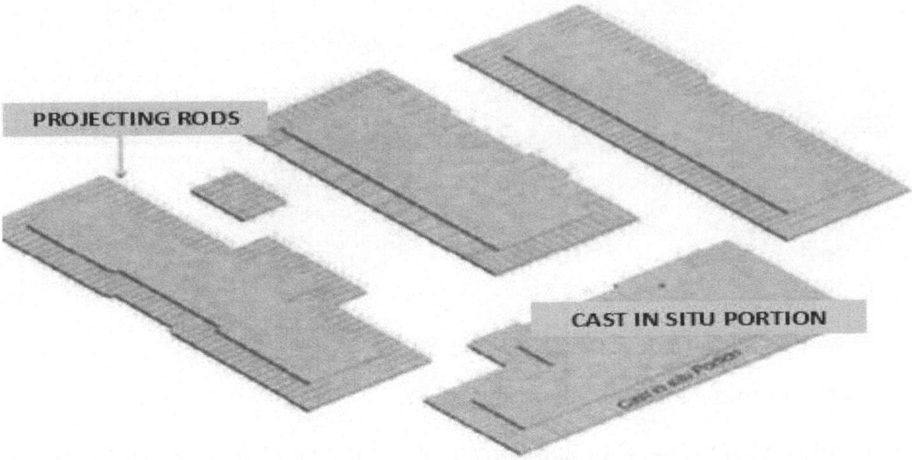

cast-in-situ concrete is a standard concrete which is poured into the specific formwork on the site and cured to get the strength of RCC elements.

Fig. 22 - Mould types - Precast Slab and Floor Panels

5. Mould types - Precast Column, Staircase and Landing:

The structural system of superstructure consists of precast construction of RCC wall, columns, slabs and beams. Floor slab shall be considered to act as a rigid diaphragm to transfer the lateral forces to walls/column. Ground floors are mostly constructed by conventional method i.e. cast-in-situ construction. It shall be designed to take the cantilever load of the above floors. First floor to higher floors shall be constructed by precast technology i.e. precast wall and solid floor slab system.

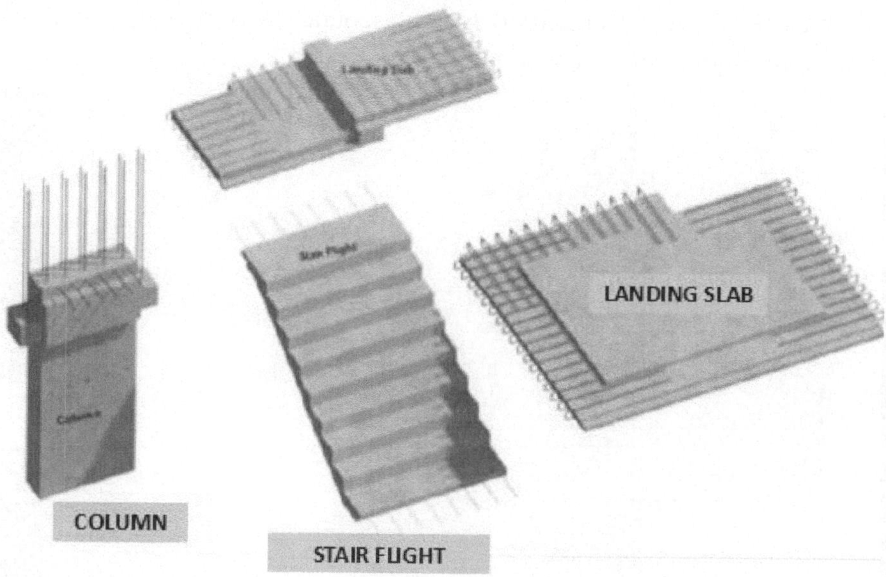

Fig. 23 - Mould Types - Precast Column, Staircase and Landing

Design & Connection Details of Precast Elements.

Precast concrete construction system has its own characteristics which influence the layout, span length, construction depth, and stability system to a great extent. In precast concrete construction, majority of structural members are manufactured in manufacturing plants away from the construction site. After that, they are delivered to the project site to be erected. Various connections have been developed to connect different types of structural elements like beam to column connection and panel to panel connection. In this manner, the imposed loads are transferred from superstructure to the foundation. Designers should consider the possibilities, restrictions, detailing, manufacturer, transport, erection and serviceability stages before finalizing the design of precast concrete structure. All loading & restraint conditions from casting to end use of structure are considered in design of precast members & connections. Connections are needed not only to transfer load but also to provide continuity and overall monolithic behavior of the entire structure. A complete system of precast elements is integrated to form a structure that behaves monolithically with sufficient strength, stiffness & durability to resist seismic & other dynamic loadings.

The connection can be classified into horizontal & vertical joints. Some typical connections details are shown.

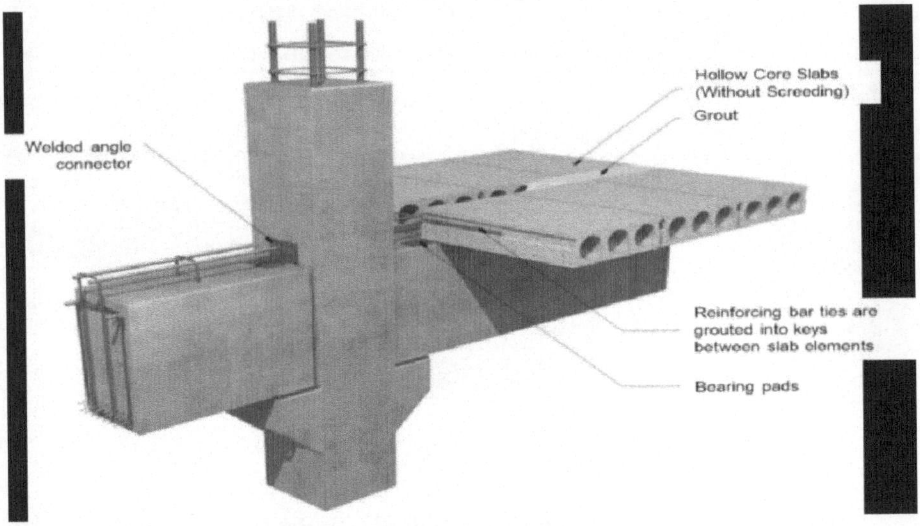

Fig.-24- Slab to Beam Connection

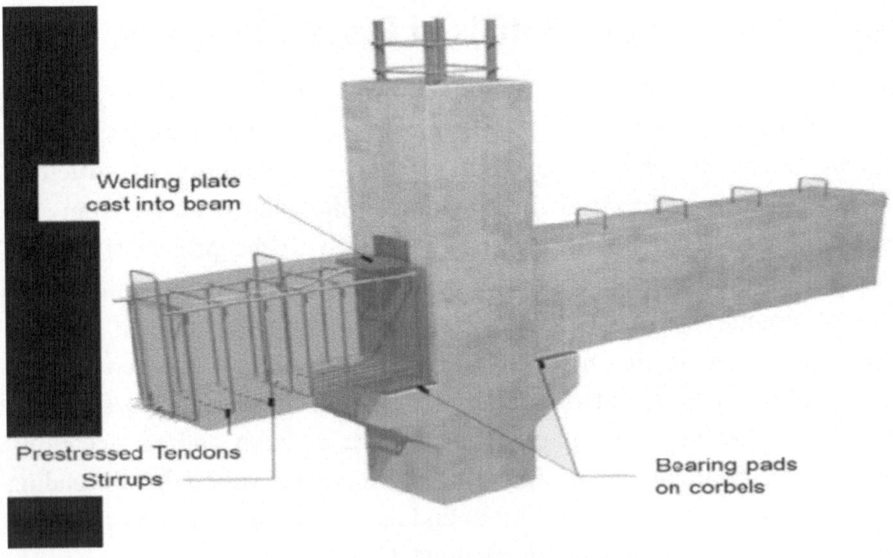

Fig.-25- Beam-Column Connection

Aim, Objective and Scope for The Study.

Aim :

As a building material, precast concrete has become widely accepted in the last several years, primarily because of cost and time reductions. The study aims to evaluate the effective use of Precast Concrete Construction in building systems in order to achieve affordable mass housing and also to minimize the intensive use of building materials.

Objective and Scope for The Study:

The main objective of this Precast Technology is to find out an optimal construction technique which has a minimal cost and gives a minimum duration of time. Precast construction, if designed and executed with thorough planning, has a great potential to respond to new market demands. Precast is smart way to achieve the sustainability objectives of Green Building.

1. To determine the type of construction that would reduce the construction cost.
2. To determine the method of construction that would help to finish the construction project in shorter duration.
3. To Identify and understand, pros and cons of Precast Concrete in residential construction (mass housing).

Limitation for The Study and Specifications for the System.

Limitation For the Study :

The study will focus only in construction phase cost, time, and quality of precast concrete construction.

Specifications for the System:

1. Ordinary Portland Cement: Shall be of 43 grades as per IS 8112:1989.
2. Fine aggregate (M Sand): Shall be as per IS 383:1970 & IS 1542:1992 and 4.7 mm.

3. Coarse Aggregates: Shall be as per IS 383:1970 and of 20 mm, 40 mm size.
4. Steel reinforcement: Shall be as per IS 1786:2008, Concrete: The grade of concrete shall be M 30 and slump forwalls, floors and roofs shall be as per IS 456:2000.
5. Brick masonry: Shall be as per IS 1905:1987.
6. Solid Block work: Shall be as per IS 2185 (Part 1):1979
7. Aluminum: Shall be as per IS 733:1983
8. Glass: Shall be as per IS 2835:1987
9. Non shrunk non-metallic grout: Cement based flowable groutshall has compressive strength of 65 N/mm2, flexural strength of 9 N/mm2 at 28 days and E-modulus of 37000 N/mm2. (ASTM C 109 & ASTM C 293-79)
10. Water proofing membrane: Fiber reinforced repair mortar shallhave compressivestrength of 45 N/mm2 at 28 days and density 2250 kg/m3 (ASTM C 109).
11. Baker Rod: Closed cell polymer-based product shall have compressive strength of 0.45 kg/cm2 min. at 25% deflection, density 22 kg/m3 min. and water absorption 0.14 gm/cm3 max.
12. Corrugated sleeve: Hot dipped galvanized prime steel sheet shall be as per IS 277:2003.

Production, Installation and Transportation Machinery.

Production Machinery:

1) Steel mould for wall/beam/slab panel/staircase, 2) Batching plant, 3) Transit mixers, 4) Vibrators, 5) Concrete buckets.

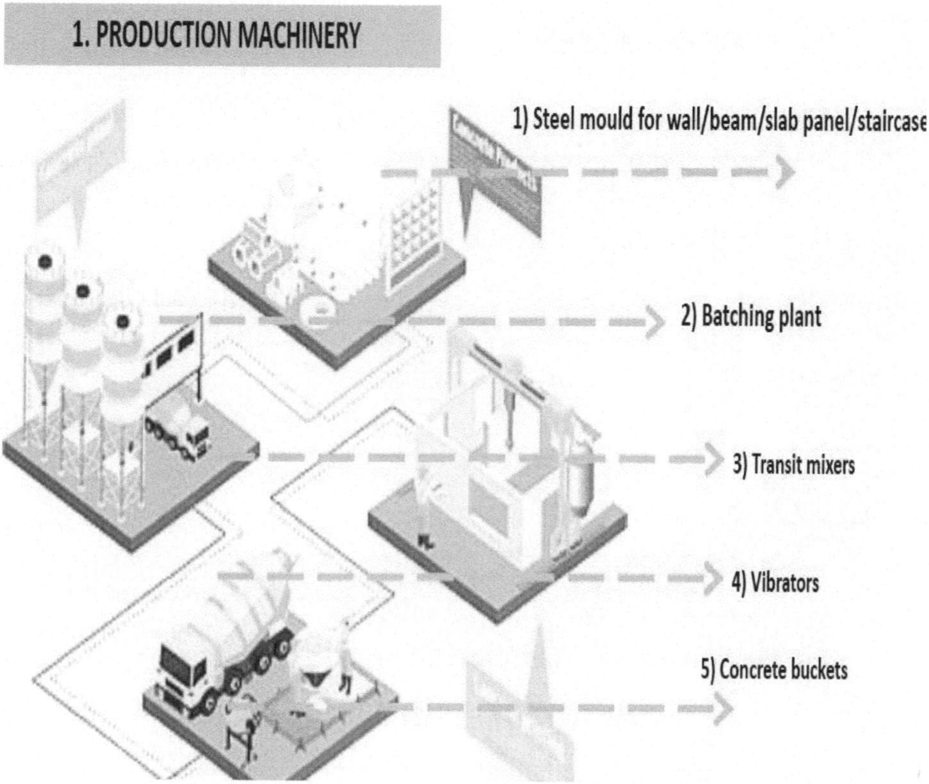

Fig. - 26

Transportation Machinery:

1) Lifting beam/ lifting clamps, 2) Ropes & lifting hooks, 3) Trailer – 20 MT capacity, 4) Wooden runner of size 40 x 60 mm/ 100 x 125 mm.

2. TRANSPORTATION MACHINERY

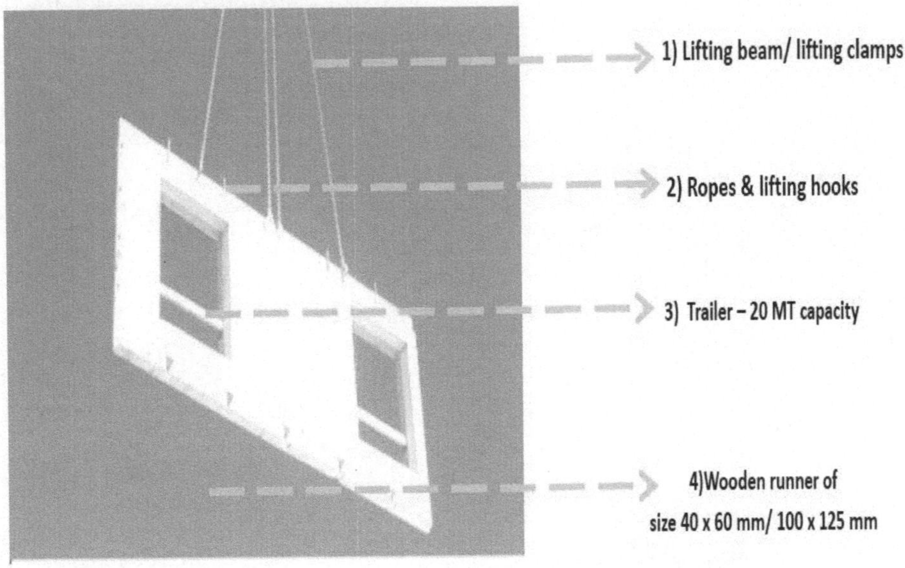

Fig. - 27

Cranes and Loaders for Installation:

1) Tower crane of 5.7 MT, 6.9 & 7.2 MT load carrying capacity and operating radius of 40 to 45 m,
2) Gantry cranes of 10 MT, 20 MT & 30 MT capacity and span of 20 to 25 m,
3) Wheel loader of 1.9 cum capacity,
4) Skid steer loader,
5) Truck mounted knuckle boom crane of 7 MT capacity.

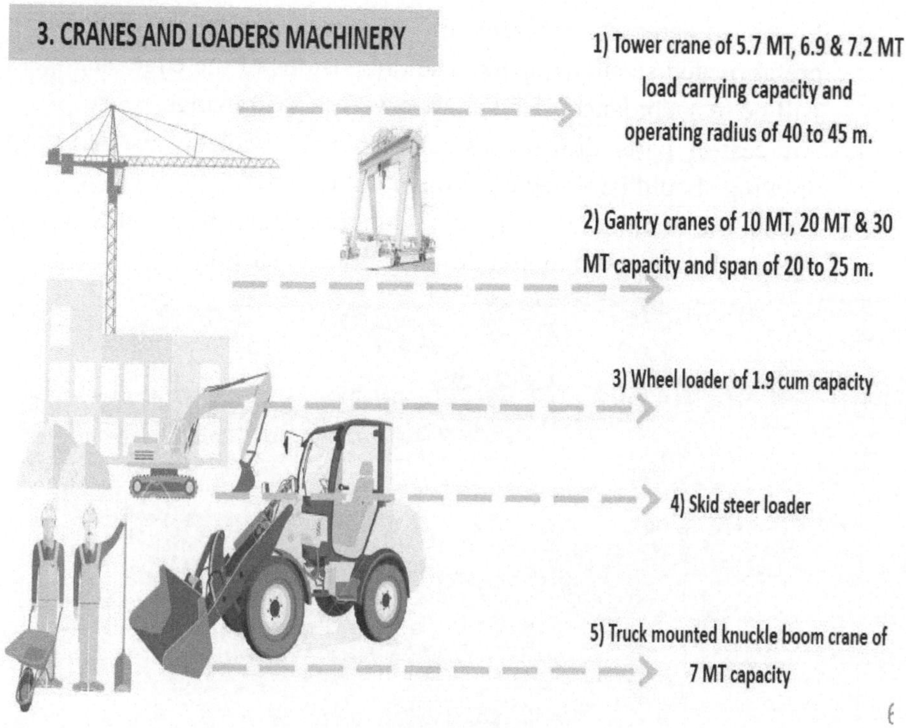

Fig. - 28

Production Procedure of Precast Concrete Elements.

A. Mould Making

B. Reinforcement Fabrication

C. Casting Procedures

D. De-Moulding and Curing

A. Mould Making:

Surface Polishing of Table Prior To Casting

1. The tables assigned for the project will be inspected for surface level using leveling instrument i.e. (Transit, Theodolite, Total station) and rectified to be within the project specified tolerances.
2. Steel moulds are then fabricated according to approved shop drawings issued by the design department.

3. Edges of precast concrete tilt elements are chamfered using prefabricated steel champers. The dimensions of the edge champers will be in accordance with the approved shop drawings.
4. All cast-in parts dimensions and location available as per shop drawing should be reflected at the mould.
5. When the mould is ready the QC inspector checks the mould against the approved mould drawing prior to casting.

Fig. - 29 - Surface Polishing of Table Prior to Casting

B. Reinforcement Fabrication:

The process of making of reinforcement mesh or cage is known as reinforcement fabrication. At present the reinforcement fabrication includes cutting, decoiling and bending, is primilarity a manual operation in

India.The proper reinforcement fabrication is very important and it must be accurate. The fabrication of reinforcement involves the following steps.Fabrication of rebars consists of cutting to length and required bending. The preparation of field placing drawings and bar lists is termed detailing. Ordinarily, the rebar supplier details, fabricates, and delivers to the site, as required. In the farwestern states, the rebar supplier also ordinarily places the bars, In the New York City area, fabrication is performed on the site by the same (union) workers who place the reinforcement. Shipping Limitations. Shipping widths or loading limits for a single bent bar and an L-shaped bar Bundles of bars occupy greater space. The limit of 7 ft 4 in has been established as an industry practice to limit the bundle size to an 8-ft maximum load width. (Manual of Standard Practice, Concrete Reinforcing Steel Institute.)

Standard Hooks. The geometry and dimensions of standard hooks that conform to the ACI 318. Building Code and industry practice are shown in Fabrication Tolerances. These are covered in Standard Specifications for Tolerances for Concrete Construction and Materials. Erection. For construction on small sites, such as high-rise buildings in metropolitan areas, delivery of materials is a major problem. Reinforcement required for each area to be concreted at one time is usually delivered separately. Usually, the only available space for storage of this reinforcing steel is the formwork in place.

Under such conditions, unloading time becomes important.

1. Proper studying the BBS and its detailing.
2. Straightening the bars.
3. Cutting the bar sizes.
4. Bending the bars according the desired shape.
5. Transporting the bars into the site.
6. Coating and treating the reinforcement.
7. Splicing the bars.
8. Binding the bars on making cage like structure.
9. The reinforcement fabrication, Tendons, and meshes add up to 20-30% the basic cost of steel.
10. All reinforcements shall pass all the test and approved as per specifications from the design department.

11. Reinforcements are the fabricated according to approved shop drawing issued by the design department.
12. When the reinforcement set up is ready, the QC inspector checks the reinforcements against the approved reinforcement drawing prior to installation to the mould.

Fig. 30 - Reinforcement Fabrication Prior to Casting

C. Casting Procedure:

- When the mould is ready for casting the QC inspector checks the mould against the approved mould drawing.
- Apply the mould release agent as thinly and as evenly as possible using fine-haired brush or roller. Any excess should be mopped up. Other application method will be in accordance with the manufacturer's instructions.

- After the reinforcement checked against the approved reinforcement drawing, the reinforcements will be placed in the mould using over cranes.
- All the cast-in parts i.e. plates, dowel bars, cast in socket, PVC pipe, recesses, etc., shall be placed in their respective location as per shop drawings.
- QC inspector checks the reinforcements and cast-in parts against the approved shop drawings prior to concrete pouring.
- Concrete are monolithically poured at the mould and compacted using poker vibrator.
- The top concrete surface will be leveled manually using hand trowel and straight edge ruler.
- When the concrete has hardened, the top surface will be checked for evenness using spirit level. Dimensional tolerances will be as QCS 2002.
- Slump test should be performed during casting of Precast Elements.
- Cube test should be performed during casting of precast elements in order to get compressive strength of concrete performed or not according to design criteria. 150mm x 150mm x 150mm in steel mould fresh concrete should be taken 5 to 6 numbers of steel moulds and put that cubes in water pounding made in laboratory and test after 7 days water curing and take compressive strength by Compression testing machine.

Fig.- 31 - Concrete Pouring for Precast Elements

E. De-Moulding and Curing:

1. When the concrete has achieved the minimum required strength 14 to 18 n/mm2 by breaking cube samples at laboratory, then the element will be demoulded.
2. Brackets are fixed at the element prior to lifting.
3. Element will be then marked with element number, project number and casting date.
4. Precast element is shifted to stockyard to achieved the 70% compressive strength for three days prior to delivery at site.
5. Any final touch up will be done during the curing period. All repair works will be as per method statement.
6. Curing of precast concrete at the tilt table can be done by either covering it with wet hessian cloth and polyethene sheet and leave it covered.

7. Application of water-based curing compound. Curing compounds play a crucial role in the production of precast concrete elements by forming a protective layer that retains moisture in the concrete mix, facilitating optimal hydration. Curing compounds are typically composed of waxes, resins, and other materials that, when applied to the surface of fresh concrete, react with calcium hydroxide (a by-product of cement hydration) to form a semi-permeable membrane. This membrane acts as a barrier to moisture loss, effectively sealing off the concrete's pores and trapping water within the mix. The preserved moisture ensures that the water-cement ratio remains constant, thus promoting continued hydration of the cement particles. Moreover, by preventing rapid moisture loss, curing compounds also mitigate the risk of premature drying shrinkage, which can lead to cracking, curling, or other forms of dimensional instability in the precast element. Precast concrete elements are cast and cured in a controlled environment before being transported to their final location and installed. Curing is an essential process for these elements for several reasons:

    - Strength development: The hydration process, in which the water in the concrete mix reacts with the cement, is responsible for the development of the concrete's strength. Curing ensures that this process can continue effectively, leading to a stronger and more durable final product.
    - Durability: Proper curing reduces the occurrence of cracks, enhances resistance to freeze-thaw cycles and de-icing salts, and improves resistance to abrasion and erosion, which are important for the longevity of the precast elements.
    - Dimensional Stability: Curing helps to improve the dimensional stability of the precast elements, reducing the risk of shrinkage or distortion, which can affect the fitting and alignment of the elements on site.
    - Quality Control: As precast elements are produced off-site in a controlled environment, proper curing is part of a quality control process to ensure the consistency and reliability of the final product. Without proper curing, variations in strength and

durability could occur, which can affect the overall performance of the construction project.

- Speed of Production: Precast concrete elements often need to be demolded and moved to make room for the next batch. Curing, particularly accelerated curing methods like steam curing, can help the elements reach their design strength faster, increasing the speed of production.

Fig. 32 - Demoulding of Precast Concrete Elements at Factory

Fig. 33 - Water Curing of Precast concrete elements after Demoulding

Delivery and Handling Procedure of Precast Concrete Elements.

1. Once the curing period is completed, according to the loadingsequence and as per the site requirements the elements will be loaded (Hollow Core Slabs, Columns, beams staircases etc. on flat bed trailers) and Precast Concrete Panels (Solid Panels, Cladding Panels and Insulated Panels on A-Frame Trailers) by overhead cranes using chain slings.
2. Wooden supports will be provided for the proper seating of the elements on the trailer. All the elements on the trailer are secured with chain slings tightened to the trailer hooks. The tightened chains coming in contact with the elements will be protected to avoid damages.
3. The loading foreman ensures that the elements loaded are in accordance with the Erection sequence and also in a safe manner. The trailer driver rechecks the load for safety conditions and follows the traffic rules and regulations.
4. The driver will be provided with the site route map and the delivery notes.
5. After reaching the site, driver contacts the Company site-in-charge. Thereafter it is the responsibility of the site-in-charge to take care of load for off-loading or erection.
6. It is the responsibility of the Main Contractor to provide proper, safe access for the trailer to the site premises.
7. Precast units will be handled and transported in a position consistent with their shape and design to avoid excessive stress which would cause damage.
8. Precast concrete shall not be moved, struck, or propped until the concrete has reached a cube strength of at least 3 times the maximum stress to which it is subjected.
9. Lift and support precast units only at designated points on the Shop Drawing. Lifting shall be carried out through designed lifting points.
10. Delivery vehicle operators shall drive with care and will be assisted by a signaler knowledgeable with traffic control procedures in areas wherein maneuvering is difficult and operators' view is obstructed.

11. Cranes supplied will be of adequate size to safely lift the precast units in off-loading from the delivery vehicles.
12. Delivery vehicles will be positioned as close to the crane unloading area as possible. It must be situated on terrain that is as level as possible.
13. Lifting Supervisor will ensure a safe working load of the hoisting equipment, rigging, and hardware before use.
14. Before mounting the precast units ensure that all safety precautions are in place.
15. Make sure all loads are properly rigged, balanced, and stable before hoisting. Tag lines must be in place and be used to control the loads and prevent rotation.

Fig. 34- Delivery of Precast Elements by Flat belt and A-Frame Trailer

Storage and Stacking of Precast Concrete Elements.

The surface of stacking area should be horizontal. The wooden runner shall be placed perpendicular to lifting points and the elements placed over runner. Number of the elements per lot should not exceed man height. In case of vertical stacking, the gap between the elements should be 150 mm to 200 mm. Stacking shall be done in such a way that slabs of longer span should be placed below that of shorter span.

a) Precast units will be stored with adequate dunnage and bracing and protect units to prevent contact with soil prevent staining and prevent cracking, distortion, warping, or other physical damage.
b) It must be stacked the right way up with the timber bearers placed just in from the ends, (maximum four times unit depth, "H") vertically above one another. Similar lengths should be stacked together.
c) Place stored units so identification marks are clearly visible and units can be inspected.

Storage of Precast Elements at Site:

Following points should be paid attention while storing precast concrete elements:

1. Elements should not be laid on a sloping ground as it could result in a abnormal loading thus resulting in overturning.
2. Ground should be inspected before placing the Element. In case of poor soil the same should be replaced with a well graded soil and proper compaction be done under supervision of qualified and experienced personnel.
3. Proper care shall be taken while placing/ removing the elements so that the adjacent element is not hit/ disturbed.
4. A warning tape should be tied around the storage area.
5. Elements are to be stacked under supervision of an experienced gang leader and to be inspected by a senior foreman for any un-symmetrical stacking.
6. At the time of stripping, ensure that concrete achieved the specified stripping strength.
7. Cubes to be tested to check the stripping strength.
8. All detachable mould parts that could prevent the elements from release to be removed.
9. Ensure that lifting clutches attached to the elements are properly fixed onto the cast-in sockets or lifting anchors/inserts.
10. Prior to stripping from the mould, the precast component to be identified with a clear and visible marking showing the element type, weight of element and date of casting.
11. The elements to be either lifted directly by crane chains or by lifting beam depending upon the type of element.
12. The elements to be lifted slowly to detach it from the mould.

13. The elements to be held by hand or rope to prevent from swinging or moving randomly.
14. After lifting the product to be stabilized for some seconds and then shifted to the required storage locations.
15. The Production personnel to inspect the product for possible defects or damage.
16. Preferably elements to be stored in a manner similar to their final position unless they are required to tilt. Tilting of elements at early stage to be avoided as far as possible. The support system used in storage is very important. A two-point, simple support system to be used to support all shapes of precast elements.

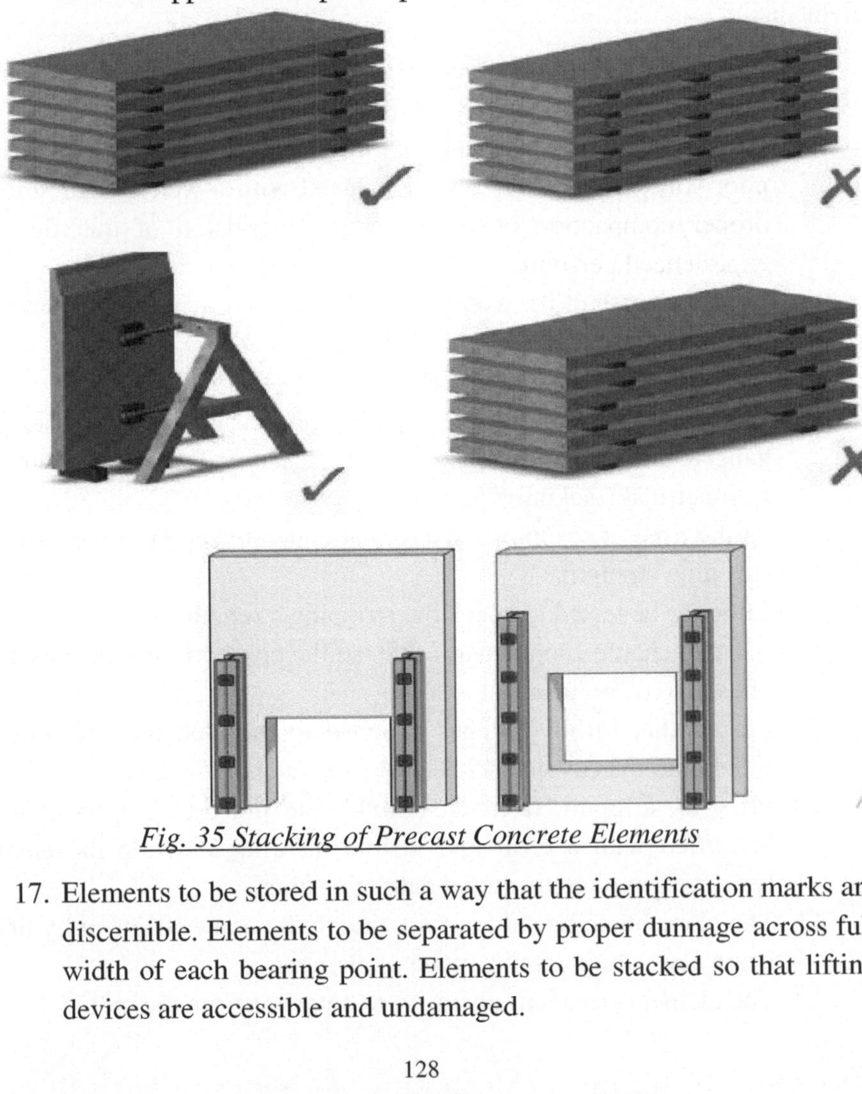

Fig. 35 Stacking of Precast Concrete Elements

17. Elements to be stored in such a way that the identification marks are discernible. Elements to be separated by proper dunnage across full width of each bearing point. Elements to be stacked so that lifting devices are accessible and undamaged.

18. Any minor defects on the surface of the precast unit to be repaired and finished as per standard methods.
19. Any major defects like chipping, broken edges, honeycombing if any, spalling, dimensional rectifications etc. to be carried out with approval of QC personnel.
20. If the precast elements are having an exposed aggregate retarded finish, it shall be transported to the washing area, where it will be placed in a rack and washed using a jet of water from a pressure pump to wash off the retarded layer to achieve the exposed finish.
21. After washing, the panel to be transferred to the covered storage area at the end of the hall.
22. During storage, the panels to be placed in the racks with the identification marks visible and will be protected against damage or staining.

Erection Procedure of Precast Concrete Elements at Site.

1 Precast Panels:

Erection of Precast Concrete Walls Panels:

Installation:

Installation shall be done by Tower or Mobile Cranes available on site. The installation crew will be divided into an unloading gang and an installation gang. The area in which precast elements are to be erected, should be cleared of people and be isolated wherever possible using safety tape. The installation process follows excavation, leveling of the ground and packing of the area, laying of the base material where the precast elements will be laid, compacting, elements' placing, anchoring using pins or packing gravel, and finally, laying adhesive or adding capstones. Precast concrete members shall be erected according to preplanned sequence. The erection plan is commonly prepared drawings If it is important for structural stability and for access to connections at specific location. The erection sequence shall avoid multiple handling of elements. Erection involves positioning, aligning and securing the components on prepared foundations to form a complete frame. Safe erection of structural construction depends on proper and timely planning. Connections between panels are made either by grout columns (hollow core slabs) or by bolting or welding plates cast with the concrete. Bolted joints should be finished by caulking with a sealant to seal the building enclosure and protect the steel connections from rusting. Key points. Precast concrete is poured into a reusable mould or 'form', cured in a controlled environment, then lifted into place. Precast concrete can be poured on site and then lifted, or poured off site and transported to site before lifting. Precast concrete can be used for floors, walls, or roofs of any building. Control points for crane signallers and riggers while performing their tasks.

2. Erection and Pick Up Technique of Precast Solid Panels:

Precast Panels horizontal members present no special handling problems once their weight and length have been considered. Weight, combined with the erection or lift off radius, will dictate the capacity of the crane required on the job. A designer should consider this and attempt to keep his maximum weights in a pre-determined range during the design stage. No erection is impossible if the owner is willing to pay for it, but in-place costs are sometimes staggering if no consideration is given to the overall erection problem. A 100- ton crane cannot erect a 100-ton piece of precast concrete. Cranes are rated by the capacity they will pick up with the shortest possible boom and at the steepest angle that they will boom up. Thus, a 100-ton mobile crane could pick up the 100 tons only with 40 ft. of boom at a radius of 12 ft. A normal working condition would be 100 ft. of boom at a 40-ft. radius. Radius is measured from the center pin or center of revolution which is approximately 10 ft. from the rear of the crane. At a working radius of 40 ft., a 100-ton crane can lift only 44 percent more than a 60- ton crane, not 67 percent more as the capacities might indicate. Tower cranes are ideal for high, multi-story erection, but their design combines a small capacity with a large radius. Gulf countries techniques of construction combine smaller precast members with cast-in-place, Four-point pickup on 54-ft. long wall panels floor-by-floor construction using tower cranes for conventional erection. Vertical members present the most difficult pick-up problems because they are hauled flat and have to be turned into a vertical position. The best way to accomplish this is with the two-line technique this usually is done with the two lines of one crane although, when the member is extra-long or heavy, it might have to be done with a second smaller crane tailing the bottom for the larger crane.

Proper planning and preparatory works shall be required before the actual installation of precast concrete elements in order to ensure quality installation. The following items shall be planned in advance:

I. Method of sequence of assembly and installation: Precast elements should be identified based on their location number and the tagged.

II. Method of providing temporary support: Elements should be supported temporarily before these get stabilized. Generally structural members with adjustable ends shall be used for securing the panels. Shims should be used to adjust the panels to ensure dimensional correctness.

III. Installation tolerances: Installation tolerances should be based on codal provisions and design considerations should be clearly indicated.

IV. Handling and rigging requirements: Elements should be checked for handling stresses before lifting and the cranes should have sufficient capacity to handle the precast panels. At least 10% impact should be considered while calculating the lifting capacity of the crane.

V. At site locations, panels shall be first unloaded and stacked or directly lifted by the crane. The element shall then be installed on the site and supported by temporary jacks. The cranes shall be released for next lifting once the temporary supports are in place. Shims shall be used to carefully align the element before grouting. The panels shall be grouted after the final adjustments are done.

3. Erection and Pick Up Technique of Precast Concrete Columns:

Columns Erection:

Column will be unloaded from the trailer or erected directly to its final position whenever possible. Otherwise, it will be stocked at firm level ground and supported by timber at the lifting points.

Erection of Precast Columns:

Anchor Bolt Connection:

1) The columns will be connected to the foundation with projecting anchor bolts anchored in the foundation.
2) Columns will be selected to the proper height using plastic shims. All principal setting out information to be provided by Surveyor.
3) The column will be braced with two bracing props fixed to the cast-in inserts in the columns and to a suitable and secure anchor drilled into the ground slab.
4) After completing the erection of columns, alignment will be done by using Theodolite, Total Station, Plumb bob and sprit level or bar level to check the verticality of columns.
5) Release the lifting tools from the column only when all fixing are fully secure.
6) When the correct position of the columns have been checked and approved, tighten the bolt. The gap between the column and foundation will be grouted by approved material.
7) Materials for erection such as grouting, and back-up rods will be as per the Engineer's approval.

4. Erection and Pick Up Technique of Precast Concrete Beams:

Erection Of Beams:

Control points for crane signallers and riggers while performing their tasks.

Beam Erection:

The most common Erection method for precast beams is with ground cranes. Cranes usually give the simplest and most rapid erection procedures with the minimum of investment, and the deck may be erected in several spans at once in relation to the availability of piers, cranes and beam delivery means.

Phases of Erection process of Beams are as follows:

1. Beams will be unloaded from the trailer and erected directly to its final position whenever possible. Otherwise, it will be stocked at site on firm level ground and supported by timber at the lifting points.
2. Any beams subjected to uneven loading will be temporarily braced with props while further erection continues. Beams will be erected after the column foundation grout has been completed.

3. Level will be adjusted by PVC shims, then the dowel location will be grouted by Nano-grout or similar approved non-shrink grout.
4. Fixing Plates are to be welded with angle plate connector with approved method and material for welding.
5. Materials for erection such as grouting and back-up rods will be as per the Engineer's approval.

5. Erection and Pick up Technique of Precast Concrete Staircase:

Erection of Staircase:

1) Stair will be delivered to site on flat bed trailers.
2) Stairs will be offloaded and stored on a firm level ground and supported by timber at the lifting points.
3) The stairs will be rigged to the crane with 4x cast in 2.5 tonnes hot-dip galvanized Frimeda anchors.
4) Level and alignment will be adjusted using PVC shims, then the dowel location will be grouted with Nano-grout or similar approved non-shrink grout.
5) After erection of stairs the gaps between the beam and the stairs will be checked for separation and equidistance, this space will later be filled with tamping grout.

6. Erection of Precast Claddings:

Precast Cladding will be unloaded from the trailer or erected directly to its final position whenever possible. Otherwise, it will be stocked at firm level ground and supported by timber at the lifting points.

A. The cladding panel will be connected to the foundation with expansion bolts.

B. After completing the Erection of Claddings, alignment will be done by using plumb bob and sprit level or bar level to check the verticality of claddings.

C. Release the lifting tools from the element only when all fixing is fully secure.

D. When the correct position of the cladding has been checked and approved, tighten the bolt. The gap between the cladding corbel and foundation will be grouted by approved material.

All the claddings are fixed with galvanized bent plate and straight plate with respect to the halfen channel and socket embedded in the precast cladding and expansion bolts on CIP beam at site. 50/30 - 150 mm rough long halfen channel is embedded in precast cladding and M16 Expansion bolt is known as a drop-in anchor It consists of a hollow sleeve that features a slotted, expandable skirt at one end, an internal expander plug, and a fastener thread at the other end is used at site for galvanized plate connection in CIP Beam. Installing a drop-in anchor is a fairly simple and quick process which requires only the anchor, a suitably sized masonry drill bit, an impact capable drill, and the anchor's expansion tool. Once the site for the anchor has been established and marked off, a hole is drilled slightly deeper than the length of the anchor. Depth of the hole is determined by making sure the nut can attach to the bolt (anchor) with enough threading remaining so that it can be torque sufficiently (usually two complete 360 the nut degree turns of). It is important to ensure that the drill bit is the correct diameter because an oversize hole will compromise the load bearing capacity of the anchor. Once the hole has been thoroughly cleared of drilling debris, the anchor is simply "dropped" into the hole with the threaded end towards the mouth of the hole. If the anchor fit is a little snug, it can be tapped into place using a mallet or hammer.Concrete cladding panels are extremely popular, especially for agricultural and industrial units. But what is concrete cladding and why has it become so common place for interior and exterior applications. Let's take a look at the reasons why you might need to use concrete cladding panels. Concrete cladding panels are made from robust and durable concrete and are applied to the inside or outside of a building to either improve the aesthetic value or to improve the building's durability. Cladding itself can be made from a variety of different materials, such as brick, stone and sometimes even wood, but concrete cladding has a much wider range of benefits than any other type of cladding material.

E. Purpose of cladding Used: The purpose of cladding is to protect a building's structure from natural elements like wind and rain but it can have other benefits, such as, insulation, noise control and it can boost the aesthetic appeal of a building.

F. How thick should be cladding: Overlap-cladding shed walls are normally 7-8mm thick, whilst tongue and groove and shiplap boards tend to be around the 12mm mark. Concrete cladding panels are made from robust and durable concrete and are applied to the inside or outside of a building to either improve the aesthetic value or to improve the building's durability.

Tolerance Limit of Precast Wall Panels Installation.

Fabrication Tolerance:

Precast concrete units must be straight and true to size and shape and comply with PCI MNL 116 and the following product tolerances:

Overall height and width of units, measured at the face exposed to view, as follows:

1. 3m or under, ±3mm.
2. 3m to 6m, to +3mm, -5mm.
3. 6m to 12m, ±6mm.

Particular Safety Precaution During Time of Erection.

1. Lifting appliances must be of adequate strength & stability, as must the load itself.
2. Lifting appliances used for lifting people must be designed to prevent falls, crushing, trapping or striking.
3. Equipment must be positioned and installed to prevent load drifting, free-falling or unintentional release from causing injury.
4. SWL and other necessary safe use information marked on the equipment.
5. Lifting operations must be properly planned, organised, supervised and carried out safely.
6. Equipment must be thoroughly examined after installation and prior to service. Equipment for lifting people must be examined and tested every 6 months; other equipment every 12 months.
7. Any defects must be reported to employer, with particulars as specified in Schedule 1 to the Regs. If defects are found, equipment must not be used until they are remedied.
8. Records must be kept.

Lifting Operations & Lifting Appliances for Erection.

<u>Cranes</u>

Types:

1.Tower Crane

2. Mobile Crane

3. Derrick Cranes

4. Gantry Cranes

5. Overhead Cranes

Safety Devices:

- SWL/Radius Indicator
- Automatic Safe Load Indicator

Factors to Consider when Selecting a Crane

- Weight & dimension of loads
- Height of lifts and distances/areas of movement offloads
- Number and frequency of lifts
- Duration of operations
- Site ground conditions
- Space available for crane access, erection, operation & dismantling

Platform Hoists

- There are 2 main types:
- Cantilever type
- Enclosed structure type

Lifting Appliances

- Jacks
- Screw Jacks
- Ratchet Jacks

• Hydraulic

• Winches

Lifting Accessories

• Ropes

• Block & Tackle

• Chains

• Eyebolts

Safe Use of Ropes& Wires

• Must be in good condition before use i.e. not frayed or corroded

• Ropes & wires must be protected from sharp edges of load by use of packing

• No knots or hitches in slings or lifting ropes

• Never drag ropes or wires over the ground

• When using more than one sling, ensure they are of the same material

Safe Use of Chains

• Never use chains with links locked or stretched

• Never hammer distorted links into position

• Never use corroded or worn chains

• Do not drag a chain from under a load

• Do not drop chains from height

• Do not use a chain over sharp corners without padding

• Chain life is increased by lubrication.

Safe Use of Lifting Accessories

• Never use hooks if there is any distortion or cracks

• Rings and shackles must be checked to see they are not distorted or cracked

Storage of Ropes, Wires, Chains & Lifting Accessories:

- Trained and competent stores
- Stores must be dry, well-ventilated and free from corrosion
- Ropes must be stored away from sunlight and heaters
- Stored material should have a systematic layout

Lifting Operations

- Site Conditions
- Access/egress routes
- Inclines/gradients
- Location of adjacent buildings
- Location of overhead lines
- Ground firm enough to support crane and outriggers
- No voids exist e.g. manholes, drains etc.
- Planning the lift
- Correct crane
- Space for access, egress and movement
- Suitable ground conditions
- Crane is level
- Underground services are protected
- No overhead or side obstructions
- Overhead power lines are isolated
- Area around lift is demarcated
- Load is within SWL of crane
- Location of lifting points ensure load is level
- Lifting points are strong enough to take the load
- There are no persons under the load at any time.
- Crane driver is given clear instruction by slingers / signaler, either by hand signals or radio

- All involved understand their role and are competent to undertake it
- All checks on crane and operator documentation are made
- Safety helmets are required to be worn by all staff are in vicinity
- Competent supervision
- Cranes must have an examination & test loading every 4 years
- A thorough examination is required every 12 months and an inspection and a test of the safe load indicator should be undertaken each week Inspections must be recorded.

Grouting Application for Precast Members.

Introduction:

The following Method Statement for Grouting is applicable to grouting of Precast panel joints, dowel bar holes and grouting of Hollow core slab joints.

Test Plan

As per project requirement, 1 set cubes of concrete samples will be taken for every 300m³ batch and will be tested at an approved independent testing laboratory. Consultants will be notified of concrete testing in an independent laboratory at least 48 hours before the date of testing.

Specifications

As per Project Specifications, grout should have minimum cube strength at 28 days of 35 N/mm². The amount of water to be added for every 25 kg. pack is between 3.5 liters to 4.5 liters.

Grouting Methodology:

1. Enough supply of grout should be available at site for continuous fill and to complete the work.
2. Clean the areas to be tamped by removing all loose particles or duct by air blower.
3. Wet the areas where the tamping grout will be applied.
4. Place the approved backing rod material between the elements.
5. Clean the grouting mixer by flushing of water before using.

6. Clean the mixer of any hardened concrete from last usage.
7. Place the pre-mixed, pre-packed grout material according to the design mix proportion specified by the manufacturer in the grout mixer.
8. Add water according to the approved consistency as per manufacturer's instruction.
9. Mix water and grout as per manufacturer's suggested length of time f. or mixing.
10. Pour the grout mix between the elements from the top. The grout must be compacted well and should be trowelled in order to be flush to the adjacent concrete.
11. All materials shall be stored in the shaded area and water below 20º C shall be used for mixing. During hot weather concreting, cold water will be used. The nano grout will be stored as per manufacturer's instruction.
12. The grout will be cured in the same manner as concrete. Curing can be done either by water curing or by application of curing compound. For more details, please refer to Appendix for technical data of approved nano grout.
13. Deep recesses formed during the grouting of element joints will be packed with finishing Material.

Equipment For Grout Mixing:

1. Cement mixer,
2. Mixing drill.

Particular Safety Precautions During Time of Grouting:

Since grout is powder containing Portland cement and sand, grout may cause irritation to skin and eyes. In case of accidental contact on the eyes, flush with water for at least 10 minutes and seek medical advice. Grout should be non-flammable.

Crack Repair Methodology of Precast Elements.

Repair of Damaged Precast Elements (if any)

QC inspector will conduct the inspection. Standard identification of defects and conducting repairs if applicable, only approved material will be used for repair work by following procedures recommended by the manufacturer.

a) Repairs to be carried out either in the Factory or at the site.
b) Inspection will be carried out to check the repairable concrete elements. Subject to repair elements will be kept in separate

designated areas. Prior to starting repair works, it will be verified that elements surfaces are free of dirt, grease, oil, curing compound, paints, and all loose material.

c) It will be confirmed that the concrete has been cured enough to achieve design strength.

d) The surface will be roughened by chipping, sandblasting, or other mechanical means to assure the bond of the repair mortar to the existing concrete.

e) Prior to conducting the concrete repair/patching, the defect/damaged precast elements shall be inspected by the AEC inspector for acceptance to proceed with the repair activity.

Substrate Preparation :

a) Saw cut perimeter maximum 10mm depth but do not cut steel.

b) Remove poor-quality concrete in the repair area until sound concrete appears. Ensure the surface is left in rough condition. All edges ideally should be cut at right angles.

c) The concrete steel reinforcement to receive the repair must be cleaned. If half of the diameter of the bar is exposed, chip out behind the rebar minimum of 10mm.

d) Areas to be repaired will be determined by visual inspection and hammer sounding. Highlight all areas where concrete quality is suspected.

Repair Material: NITOBOND EP AND NITOFILL EPLV:

Major Plant and Equipment:

For the execution of the works, suitable equipment and tools will be utilized as follows Appropriate:

1. cleaning cloths

2. Grinder

3. Necessary buckets of clean water

4. Electric drill

5. Steel brush

6. Compressed Airline

7.Plastic Foam

8. Mixing Paddles

9. Mixing Buckets

10. Trowel or spatula

11. Plywood

12. Plate Tamper (Hand Tool)

13. Epoxy Injection Pump

14. Diamond Cutter

15. Chipping Gun

List of Appendices:

1. (RENDEROC LAXTRA PLUS)

2. (NITOBOND EP)

3. (CONBEXTRA GP)

4. (NITOFILL)

5. (SIKAREP-100)

Acceptance and Rejection Criteria of Precast Elements.

Scope:

Following outline criteria shall be used to determine whether the element should be repaired or rejected:

A. Rejection Criteria:

Elements will be rejected in the event of any of the following faults being evident.

i. Structural:
   a. Excessive structural damage that cannot be repaired.
   b. Exposure of reinforcement that cannot be corrected.
   c. Units so dimensionally incorrect that cannot be repaired.

ii. Architectural:

The range of acceptability with respect to colour and texture of architectural finish shall be evaluated when viewed from a distance of not less than 6.1 Meters considered as the normal viewing distance of the structure.

a. Very poor aggregate exposure beyond repair
b. Large variance in color or exposed aggregate which cannot be corrected.

B. Repair Criteria:

I. General:
   a. Repairable cracks.
   b. Poorly consolidated concrete (e.g. honey comb) provided the structural integrity is not affected to an acceptable limit.
   c. Damaged lifting hook area, corners and edges. These repairs are, if possible, left until the element is erected to save a recurrence of damage in delivery and erection.
   d. Cast in items misalignments.
   e. Blocked electrical conduits shall be replaced without affecting the structural integrity.
   f. Missing dowel bars.

II. Architectural:
   a. Sharpening and straightening of chambers and joints between fair faced and architectural finishes.
   b. Obvious variance in the exposed aggregates or colour when viewed from a distance of 6.1 meters.
   c. surface indentations and blow holes.
   d. Replacing cast in Architectural features which are either broken or misplaced.
   e. Minor cracks within exposed face or other faces.
   f. Colour variations in the external face of the element.
   g. Any other work that can arise during the project execution but which can be repaired to retain the architectural finish but without affecting the structural integrity.

Waterproofing.

External joints shall be sealed with baker rods and sealants after filling the joints with grout to avoid the leakage. Additional waterproofing treatment shall be provided at external joints and wet areas to ensure water tightness.

Mechanical, Electrical & Plumbing Fitting.

Mechanical, electrical & plumbing fittings shall be kept open or concealed as per the requirements. For concealed fittings, provision for grooves, blackouts shall be made in casting moulds. The conduits and electrical boxes shall be embedded and fixed in moulds before casting. For open fittings, these shall be fixed after erection at site. For firefighting systems, provision of National Building Code (NBC) and local body law shall be adhered to. Precast concrete shall be designed for fire rating of 1 to 2 hours based on codal requirements. Minimum precast concrete wall thickness of 120 mm shall be provided for 1 hour fire rating as per IS 456:2000. Variety of shapes, colours, textures and finishes may be obtained with precast concrete. The surface treatments shall be done by rebating, grooving, surface coatings, cement-based renders, oxide coloring etc. Precast concrete facades of various shapes, colours and textures may be moulded and installed.

Connection Details.

The connection were simple and special dowel bars, corbels and provisions for connection were planned in advance for the members as per the design. The joints were made monolithic by grouting them with non-shrink grout. Onsite, wet grouted joints for horizontal connections between wall to wall, slabs to wall, termed as stitched joints were selected.

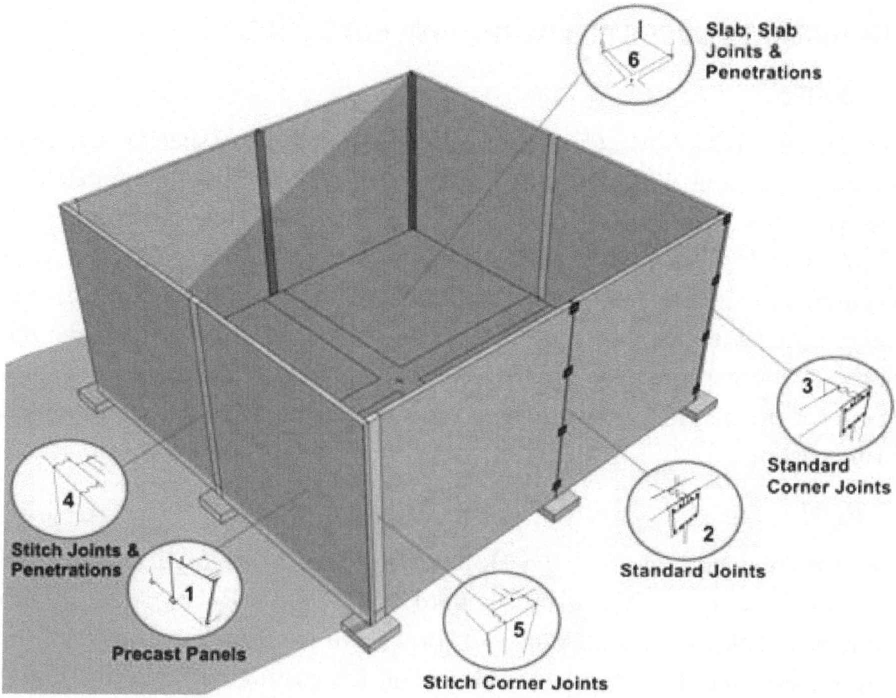

Fig. 36 Connection Details

Introduction of Hollow core Slabs.

A hollow core slab, also known as a voided slab, hollow core plank or simply a concrete plank is a precast slab of pre-stressed concrete typically used in the construction of floors in multi-story apartment buildings. Slabs in pre-stressed concrete are usually produced in lengths of up to 200 meters. Pre-stressed Hollow core floor units have longitudinal cores of which the main purpose is to reduce the weight of the floor. The elements are available in different depths in order to satisfy the various performance needs for span and loading. After the customer had placed an order for hollow core slabs with precast company. Precast supplier company will comment on the design from the client or make additional proposal to the customer. All relevant structural drawings in its latest revision and loading details should be provided for pre-stressed precast hollow core supplier reference. Any ducts or openings larger than 150 mm should be finalized during design stage. Production of Hollow core slabs can begin after design has been finalized.

Production Procedure of Hollow core Slabs.

1. Casting

The casting ofHollow CoreSlabs is doneby using aspecial concrete extruding machine. The extruding of concrete takes placewhile the machine is moving on top of a long casting steel bed. The bed is 150 meter long and 1.2 meter wide. While the machine is moving itreceives a dry concrete with minimum concretestrength of 60 N/mm2. The machine transforms the concrete into the shape of the Hollow core slab and at the same time the concrete is compacted by shear vibration technique. There is no mould except for the bed itself. To roughen the surface, the slabs will be broomed during casting.

2. Reinforcement

The reinforcements used in the Hollow Core Slabs are high tensile steel strands complying with British Standard BS 5896/3 - Relax Code 2, with diameters of 9.3 mm, 12.5 mm and /or 15.7 mm. Prior to casting of the Hollow core slab, the strands are placed on the casting bed over the entire length, the strands are pre-tensioned. The exact positions of the strands are secured by strand-guides inthe machine at each end of the casting bed. The number ofstrands used in the Hollow core slab depends on the design requirements.

3. Openings and Cut-Outs

Larger openings are usually cut in the wet concrete immediately after casting, while smaller openings are drilled using a core drilling machine.Openings of 150mm and smaller are made after the concrete has hardened at the construction site. All openings drilled on site shall be coordinated with design department for approval before drilling. For openings bigger than 150mm, Designer will provide the location and details of all openings and cutouts prior to fabrication. Openings and cutouts should be clearly indicated in the approved productionshop drawings. Factory will produceall HCS according to approved production shop drawings. Factory will provide tendon profile along x and y axis with reference to the edge ofHCS to avoid cutting the tendons during on-site core drilling. If the width of an opening inthe floor exceeds 800mm, it is most advisable to cut the slab completely through.The slab isthen suspended from the adjoining slabs by a steel saddle beam.As we know that PC(Precast Concrete) strand is most important part of HC slab as this is where the strength comes from, so without approval of Structural/design engineer cutting strand at site is not acceptable as it would jeopardize structural integrity of buildings.

4. <u>Curing of The Hollow core Slab</u>

Immediately after casting, the Hollow core slab is covered by an insulating rubber carpet ensuring that the slab is kept humid and warm for effective curing until the slab is being cut as soon as required concrete strength is reached which will be determined by breaking cube samples at laboratory. The cube testing will be done for each hollow core bed. After cutting of the slab, the slab is lifted & transported to the storage yard or to site stockyard. In case of early transport to site, curing compound will be applied according to manufacturer's manual. The curing compound shall be water-based and should be approved by the Consultant prior to use.

5. <u>Cutting of the Hollow core Slab</u>

After curing on the casting bed for approximately 10 - 16 hours, the concrete has achieved minimum concrete strength to transfer the pre-stressing forces from the strands when the strands are released / cut. If the concrete has the required strength of at least 30 N/mm2 as per cube testing at laboratory, the slabs are cut as per design length by using a special cutting machine with diamond blade.

6. <u>Storage of Hollow core Slabs</u>

After control the slabs are transported to the storage yard or site stockyard where the slabs are stacked and covered with wet Hessian or polyethylene sheet or curing compound is applied. Under this cover the slabs arekept wet and humid for curing up to 7 days from the date of casting or as per British Standard minimum requirements. Hereafter the cover is removed and the slabs are ready for delivery.The slabs are stacked and separated by timber at each end maximum 500 mm from slab end. Elements should not be laid on a sloping ground as it could result in an abnormal loading thus resulting in overturning. Ground should be inspected before placing the Element. In case of poor soil, the same should be replaced with a well graded soil and proper compaction be done under supervision of qualified and experienced personnel. Proper care shall be taken while placing/ removing the elements so that the adjacent element is not hit/ disturbed. A warning tape should be tied around the storage area. Elements are to be stacked under supervision of an experienced gang leader and to be inspected by a senior foreman for any un-symmetrical stacking.

Delivery Procedure of Hollow core Slabs.

Once the curing period is completed, according to the loading sequence and as per the site requirements the elements will be loaded (Hollow Core Slabs on flat bed trailers) by overhead cranes using chain slings. Wooden supports will be provided for the proper seating of the elements on the trailer. All the elements on the trailer are secured with chain slings tightened to the trailer hooks. The tightened chains coming in contact with the elements will be protected to avoid damages. The loading foreman ensures that the elements loaded are in accordance with the Erection sequence and also in a safe manner. The trailer driver rechecks the load for safety conditions and follows the traffic rules and regulations.

Erection Procedures of Hollow core Slab.

The HCS will be lifted and placed in to the position on the structure as per the approved layout Plan. Level and position will be checked by the Site Engineer and necessary adjustments will be adopted. Slabs are lifted directly from the trailer and in to position.Full slabs are lifted using the specially designed lifting equipment or using textile sling belts. The lifting clamps shall be placed 500 mm from the ends of the slabs. The chains between clamps and beams shall be 90°. The angle between the chains on the lifting shall be less than 60°.This rule will be observed to make sure the cantilever part of the Hollow core Slab does not break and the lifting beams does not collapse.When lifting narrow slabs (width less than 1200mm) & slabs with cut-outs near the end, sling belts will be used. Also, full width slabs can be lifted by means of sling belts.When lifting with clamps, always use the safety chains underneath the slab. The chains should not be released until the slab is about 100mm from the support. Hollow Core Slabs shall be erected on shims leaving a gap of approximately 20mm between the soffit of the slab and the top of the Precast wall panel.Final leveling of the Hollow Core Slabs by adjusting the shims, after lifting the slab with hydraulic props.The site operation will erect, in most cases, as an average for 150mm-265mm thick Hollow Core Slabs 8-10 units per hour and for 320mm-500mm thick Hollow Core Slabs 4-6 units per hour depending on the difficulty of the job.

Erection of Hollow Core Slabs on Shims.

Hollow core slabs shall be erected on leveling shims for the following reasons:

Aesthetical Reason

Shims are used for leveling the slabs so the soffit of the individual slab is in the same level after erection.

Structural Reason

The evenness of the contact zone along the support is essential for the bearing capacity of the Hollow Core Slabs, as the shear capacity depends on full support over the 1.2m width.

Two (02) nos. of shims are used for initial support and leveling at each end of the slab, as shown on the sketch below.After the final leveling, the slabs shall be tamped with a relative dry sand cement mortar with fck≥35 Mpa, which will be the permanent support of our slabs.It is important that the shims are made of a compressive material to ensure that the slabs will be resting 100% on the width tamped mortar bed and not on the temporary 75mm width shims.

Problems to Be Avoided after erection.

If grouting and tamping, and curing of this, is not done immediately after the erection and leveling, or if it is done improperly, the following problems may arise:

1. If the slabs are loaded before grouting and tamping, i.e. with building materials for further upwards construction, then these loads will definitely push the slabs out of level and maybe, even cause cracks due to overloading.

2. If holes, i.e. for fan hook, electrical installation, etc. are drilled into the slabs before grouting and tamping, and then the drilling will possibly push the slabs out of level.
3. If the joints are left ungrouted for a longer period of time, then the joints will inevitably be filled with dirt and dust, which will prevent a proper grouting, and most importantly, will spoil the structural shear key between the slabs.
4. Other problems can arise. So, grout and tamp should be done immediately after erection and leveling.
5. When grouting and tamping is completed, the slabs are, in most cases, covered with 50-100mm screed or structural topping.
6. When curing the screed, it is recommended to cover the screed with wet canvas and polyethylene sheets instead of continuous pouring of water.
7. This procedure will give good curing humidity and reduce the amount of water used for curing.

Tamping Grout of Hollow core Slabs.

- Prepare the tamping grout according to the approved mix design.
- Clean the surface and remove all loose particles and any dust by air blower.
- Wet the areas that are subjected for tamping.
- Apply the tamping grout between the elements.
- The tamping grout must be compacted until it reaches halfway to the areas that are subjected for grouting.
- The tramping grout will be trowel finish in order to be flush to the adjacent element.
- The surface must be cured immediately after finishing and be done at finishing time without any delay.
- All materials shall be stored in the shaded area and water below $20^{O}C$ shall be used for mixing.

Advantages and Dis-advantages of Hollowcore Slabs.

Advantages of Hollow Core Slabs

Hollow core slabs arepre-stressed concrete elements that are generally used for flooring.Some of their effects are as follows:

- Long span, no Propping
- Flexible in Design
- Fast Construction
- Light weight structures
- Floor voids and penetrations are available
- Trimmer Beams can also be used

The slabs have between four and six longitudinal cores running through them, the primary purpose of the core being to decrease the weight, and material within the floor, yet maintain maximal strength, the slabs are reinforced with 12mm diameter steel strand, running longitudinally.Currently we offer a range of five slabs depths; 200, 220, 300-, 320-, 400- and 500-millimeters slabs. Depending on the project requirement, in particular span and loading performance, a particular slab depth is chosen. Hollow core slabs make the ceiling and the floor of a room into a mass storage of energy with wide surfaces for cooling and heating. This gives a thermal stability to the room. As these surfaces are warmer in winter and coolerin the summer compared to the room temperature, thermal comfort is increased. Circulating cold night air in the core system may provide enough cooling during day time, without mechanical cooling.

Dis-advantages of Hollow core Slabs

- Camber in Prestressed Hollow core slabs
- Very small margin for error
- Connections may be difficult
- Somewhat limited building design flexibility
- Joints between panels and Hollow core slabs are often expensive and complicated.

- Skilled workmanship is required in the application of the Hollow core slabs on site.
- Cranes are required to lift Hollow core slabs.

Case Study on use of Precast Concrete Technology for Construction of High-Rised Residential Buildings.

1. Case Study on use of Precast Technology for Construction of High-Rised Residential Buildings:

Check List: Checks are to be done mainly while erecting, transporting and manufacturing. So data has been collected regarding various checks need to be done in precast construction and a check list is prepared.

Risk Analysis: A questionnaire is done to know about the various risks and their level of impact on and off the people on the site. It is prepared based on the factors or conditions that cause injury such as due to heavy lifting, structural collapse, improper barricades and braces, working with heavy equipment etc. The questionnaire was surveyed from various people working in the field of precast.

This paper highlights the case study for adoption of Precast Technology to achieve fast-track, sustainable, and cost-effective construction of high-rise buildings in Indian Scenario.

A. Real-Time Case Study- Dream Valley project is located in Greater Noida (West), Delhi NCR, India. It's a residential township with total 47 high-rise residential towers, 379 villas, commercial & institutional building, and other developments. The total built-up area of project is more than 10 million Sqft.The high-rise residential towers are divided in to six series from A to F with further classification as A1 to A7, B1 to B6, C1 to C12, D1 to D2, E1 to E8, and F1 to F12. After a lot of brainstorming and feasibility study, Series A, D, E & F were planned to be constructed by using precast construction technology. To make the case easy to understand, this study will discuss the construction of series A having total seven towers. As of today, all the seven towers are nearing completion. There are 12 apartments per floor with carpet area of about 430 Sqft. for 1BHK. All apartments at of series A are identical that made precast technology a viable option.

Dream Valley – Series A (Tower A1-A7), High Rise Residential Buildings	
Overall built-up Area (Precast+CIS)	10 Million Sqft
Built-up area planned in Precast	3.79 million Sqft
Built-up area of A series (A1-A7)	8.82 lacs Sqft
Structural frames	2B+G+18 Floors
Total apartments	1500 apartments; 12 apartments per floor
Structural system (Sub-structure)	Raft foundation (CIS)
Structural system (Super-structure)	Hybrid Precast Construction (Shear walls & Core in CIS)
Basement, Ground & terrace work	Cast In Situ
1st to 18th floor	Precast beams, wall panels, balconies, claddings, staircase, & hollow core slabs
Total precast concrete volume (m3)	~16000
Total Nos. of Precast elements (Tilts + HCS)	25000

Challenges Experienced:

Following are some of the key learning challenges which were encountered on the way.

1. As the project was located in seismic zone IV, the suitable design was adopted.
2. Lack of skilled manpower with the knowledge of precast industry, operation of various machineries & erection, which was overcome by providing detailed training.
3. It was difficult to counter the typical mind-set of people to adapt to change in construction technology, patience & willingness to experiment with a new technology.

   Although, European and other standards were available for ready reference but unavailability of IS codes was deemed important. The designers from India had only few Indian Standard to refer. The knowledge and understanding acquired during this case were carried forward to the projects underway. Many improvisations were based on the experiences of this projects.

RESULTS ACHIEVED:

Some astonishing results along with learning experiences achieved through this landmark project are noted as follows:

1. 7 buildings of 2B+G+18 floors were completed in 18 months of timeline was made possible through precast technology.
2. Achieved a slab cycle of 10-12 days for a slab area of 7000 Sqft.
3. Superior quality & finish of construction.
4. Cost & time optimization with the use of hollow-core slabs.
5. Reduction in concrete & steel factor per Sqft of built-up area.
6. Elimination of brick work and plaster by the use of precast wall panels & cladding.
7. Automation & mechanization of construction project improved overall productivity.
8. Improved site safety & considerable reduction in wastage, dust & noise on site, thus reduced ecological footprint of project.

Inference:

The precast concrete technology has already arrived in India, due to large size projects, need for quality construction with speed & reduced labour force. All these advantages can be exploited to the maximum by careful planning & designing. This case study was aimed to share the hands-on-experience, and to aware the people about the potential the precast offers. From the above case study, it can be concluded that precast construction, if designed and executed with thorough planning, has a great potential to respond to new market demands. Precast is smart way to achieve the sustainability objectives of Green Building.

CONCLUSIONS

1. Summary

The precast concrete technology has already arrived in India, due to large size projects, need for quality construction with speed & reduced labour force. All these advantages can be exploited to the maximum by careful planning & designing. This case study was aimed to share the hands-on-experience, and to aware the people about the potential the precast offers. Following learnings could be the key takeaways though this case study.

Hybrid precast approach offers more benefits in terms of time, cost, and quality. Repetition of precast components is essential in order to meet quantity for cost effectiveness. Improved quality control, reduction of construction time, construction waste, dust & noise on site, and labour requirement on site. it results in higher useful gross floor area which contributed to significant cost benefits. A considerable reduction in steel factor can be achieved even for seismic zones. Precast technology allowed for flexible design. Precast is smart way to achieve the sustainability objectives of Green Building.

Conclusions

1. From the above case study, it can be concluded that precast construction, if designed and executed with thorough planning, has a great potential to respond to new market demands.
2. Adopting mix of CIS &factory-made precast units wherever required, instead of usually considered 'all precast or no precast' approach offers more benefits in terms of time, cost, and quality.
3. Repetition of precast components is essential in order to meet quantity for cost effectiveness.
4. The adoption of precast compared with traditional construction demonstrated significant advantages, such as improved quality control, reduction of construction time, construction waste, dust & noise on site, and labour requirement on site. In addition, it results in higher useful gross floor area which contributed to significant cost benefits.
5. A considerable reduction in steel factor can be achieved even for seismic zones IV & V.
6. It becomes easy to perform non-destructive testing (NDT) if the need arises and becomes easy to mitigate.
7. Precast has evolved towards the use of non-standard design approach with modular elements optimising site opportunities & constraints.
8. For commercial construction and other standard products like boundary wall etc. precast has unprecedented benefits.
9. Precast technology allowed for flexible design and longer clear spans in the non-tower areas like parking.

10. Precast is smart way to achieve the sustainability objectives of Green Building.

2. Scope for Future Work

    1. To analyse and to study of Precast Concrete in construction management aspect in terms of cost and time.
    2. To analyse the cost of structure due to savings in materials.
    3. To familiarize with the Precast Concrete construction methodology with due focus on the difference from the conventional construction.
    4. To Identify and understand, pros and cons of Precast Concrete in residential construction (mass housing).

3. Scope for Services

Client Representation, Program Planning, Design Team Management, Time Scheduling, Project Documents Review, On-site Contractors Co-ordination, Quality Control, Health & Safety Provisions Co-ordination and Ensuring all necessary Compliance.

Introduction about why buildings constructed by using Precast concrete Technology is considered as Eco-Friendly Building Construction.

Precast Concrete is a very sustainable building material with regard to the energy consumption during its production and its characteristics during its useful life. Concrete is a natural building material and 100% recyclable. Concrete consists of water, sand, gravel and cement. Water, sand and gravel naturally exist in our environment. For the production of cement, the raw materials limestone, clay, sand and iron ore are kilned, gypsum is added and afterwards the mixture is finely ground. At the production of precast concrete in the plant, the recycling of the fresh concrete which has not been used is a big priority. Fresh concrete is concrete which is not yet cured. The fresh concrete which has not been used is returned into the production cycle and can therefore be recycled to almost 100%. Reduced consumption of energy, water and building material.

Modern precast concrete plants are controlled and monitored by a master computer. The master computer exactly calculates the necessary amounts of cement, construction steel, water etc. and therefore allows for an optimal utilization of materials not only during the production of precast concrete elements are savings. The positive characteristics of the building material also take effect during the utilization of a precast concrete building. Minimum waste output in the production process and reduction of packaging material. The use of such technology helps in saving of the time taken for similar projects using normal construction methods and technology. Precast concrete is the new method concreting used all over the world extensively. It is also used in India as well but not very often. In this method concrete is prepared, poured in formwork and cured in controlled environment. In this method quality of concrete can be assured because concrete is prepared and cured in controlled environment. Precast construction system is generally a large panel system, modular system or a combination of both. Precast Large Construction Panel (PLCP) system consists of various precast elements such as walls, beams, slabs, columns, staircase, landing and some customized elements that are standardized and designed for stability, durability and structural integrity of the building. Precast residential building construction involves design, strategic yard planning, lifting, handling and transportation of precast elements. This technology is suitable for construction of high-rise buildings resisting seismic and wind induced lateral loads along with gravity loads. The building framing is planned in such a way that maximum number of repetitions of moulds is obtained. These elements are cast in a controlled factory condition. The factory is developed at or near the site which provides an economical solution in terms of storage and transportation. Two main types of precast concrete elements, namely precast reinforced concrete elements and precast pre-stressed concrete elements are used as per the details given below: *Precast concrete elements* – Concrete components of a building prefabricated in precast yard or site and shall be installed in the building during construction. Precast pre-stressed concrete elements. These shall consist of pre-stressing tendons with in the elements to provide a predetermined force needed to resist external loadings and cracks such as hollow core slabs, beams and planks.

(I) Mass- Precast concrete products can act as effective barriers to vehicular traffic due to their size and weight. In the current world climate, precast concrete products such as planters, bollards and highway barriers increasingly are being used to provide protection for a wide variety of venues.

(II) Chemical Resistance- Precast is resistant to most substances. However, no material is completely immune to attack form aggressive chemical agents. Thus, it is wise to choose the material with the longest expected service life Precast concrete products can be designed to withstand anticipated corrosive agents.

(III) Quality Control- Because precast concrete products typically are produced in a controlled environment, they exhibit high quality and uniformity. Variables affecting quality typically found on a jobsite-temperature, humidity material quality craftsmanship-are nearly eliminated in a plant environment.

(IV) UV Sensitivity- Unlike some other materials, precast concrete does not degrade from exposure to sunlight. This is extremely beneficial for above-ground applications.

(V) Environmentally Friendly - After water, concrete is the most frequently used material on earth, it is nontoxic, environmentally safe and composed of natural materials. Buried throughout the world, precast concrete products help convey water contributing to poor water quality.

(VI) Weather Resistance- Precast concrete is well-suited for exposures to all types of weather conditions. Reduced Weather Dependency - Precast increases efficiency because weather will not delay production. In additional weather conditions at the jobsite do not significantly affect the schedule. This is because it requires less time to install precast compared with other construction methods, such as cast-in-place concrete. Precast concrete can be easily installed on demand and immediately backfilled-there is no need to wait for it to cure.

(VII) Water Tightness - Standard watertight sealants are specially formulated to adhere to precast concrete, Making watertight multiple-seam precast concrete structures possible.

(VIII) Ease of Installation- Since precast products are designed and manufactured for simple connection, many components can be installed in a short time.

(IX) Modularity - Because of the modular nature of many precast concrete products, structures of systems of nearly any size can be installed in a short time.

(X) Efficiency- Precast concrete products arrive at the jobsite ready to install. There is no need to order raw materials such as reinforcing steel and concrete, and there is no need to spend time setting up forms, placing concrete or waiting for the concrete to cure.

(XI) Aesthetics - Precast concrete products are both functional and decorative. They can be shaped and molded into an endless array of size and configurations Precast concrete can also be produced in virtually any color and a wide variety of finishes (acid-etched. stand blasted, smooth-as-cast. exposed-aggregate) to achieve the desired appearance f or building and site application.

(XII) Low Maintenance- Precast concrete requires or on maintenance, which makes it an ideal choice for nearly any design solution. Fire Resistance- Precast concrete is noncombustible, also, concrete does not its structural capacity nearly as quickly as steel. Other materials besides concrete and steel are flammable and / or do not perform well in elevated temperature.

(XIII) Fire Resistance- Precast concrete is noncombustible, also, concrete does not its structural capacity nearly as quickly as steel. Other materials besides concrete and steel are flammable and / or do not perform well in elevated temperature.

Quality Control Procedure of Precast Concrete Members.

Quality control specialists are responsible for inspecting and/or testing products before, during, and after the manufacturing process to ensure that products align with company and legal standards. The professionals monitor nearly every manufactured product, ranging from food and clothing to glassware and cars.

Quality control forms an integral part of precast production. An efficient quality system is most critical in the mass production of precast components in any project. In this study, the quality control system implemented in a precast factory is discussed. The precast factory is set up for the manufacture of large quantity of tunnel segments under a contract. Processes in the precast manufacture are discussed with respect to the control procedures in the quality inspection plan. The standard tests involved, roles of inspectors and corrective actions on-site are highlighted. Critical issues pertaining to the productivity and quality of precast production are reviewed. These include the quality of staff and maintenance, which could potentially affect the efficiency of the quality system. Some considerations should also be given to improve the workflow and productivity.

Quality control forms an integral part of precast production. An efficient quality system is most critical in the mass production of precast components in any project. In this study, the quality control system implemented in a precast factory is discussed. The precast factory is set up for the manufacture of large quantity of tunnel segments under a contract. Processes in the precast manufacture are discussed with respect to the control procedures in the quality inspection plan. The standard tests involved, roles of inspectors and corrective actions on-site are highlighted. Critical issues pertaining to the productivity and quality of precast production are reviewed. These include the quality of staff and maintenance, which could potentially affect the efficiency of the quality. The important purpose of Quality Control is to ensure that all work performed will be in accordance with the specifications. The main objectives are listed as,

- To monitor and control the production in all stages and deter the occurrences of errors so they can be corrected before they become a damaging and costly phenomenon.
- To obtain the highest possible standard of precast product quality as demanded by relevant standards and specifications
- To guarantee by documentations that the quality of the precast products has the desired level according to functionality, durability, and aesthetic appearance.
- To highlight errors not discovered during the production and control process.

The following are some of the most important aspects in Quality Control:

- Management commitment to quality and quality control programs
- Qualified personnel for all stages of design and production
- Testing and inspection of various materials put to use
- Clear and complete shop drawings
- Accurate stressing procedures for prestressed elements
- Control of dimensions and procedures
- Proper positioning of all embedded items
- Correct size and positioning of reinforcement steel
- Proper proportioning and adequate mixing of concrete
- Proper handling, placing and consolidation of concrete
- Appropriate and controlled curing
- Proper procedures for detensioning and stripping
- Proper procedures for handling, storing, transporting and erecting precast members
- Complete and accurate production documentation

Quality Control Procedure:

In general, the quality control procedure for precast concrete products includes the following,

- Identification, examination, testing and acceptance of materials.
- Inspection of beds/tables/moulds and forms prior to casting
- Checking the dimensions of elements, reinforcement steel, inserts, other incorporated materials, openings, blackouts etc.
- Inspection/test of concrete mixes, cube tests and production samples.
- Regular inspection of batching, mixing, conveying, placing, compacting, finishing and curing of concrete.

- Final inspections of finished precast product prior to shipment i.e. monitoring dimensions, bowing, blackouts, and finishing.
- General observation of plant equipments, working conditions, weather and other items that may potentially affect the product.

The QA/QC team is responsible for the Production yard and Site quality control of precast elements through the following activities.

- Inspection of raw materials
- Inspection of precast mould fabrication
- Inspection of reinforcement and mould ready for casting
- Inspection of precast elements in the process of production
- Inspection of finished precast elements
- Final inspection of precast elements
- Curing of precast elements
- Inspection of repairs and rejection of precast elements
- Final inspection of erected precast elements

Quality assurance and quality control of precast concrete members are based on two levels

- In-house quality assurance programme with continuous in house control
- Plant certification with quality control supervised by an independent body

<u>Quality Control Testing Equipments:</u>

Minimum Laboratory equipment should include following

- Scale , 25kg capacity and accuracy of 50 grams
- 2kg capacity and accuracy of 0.1 grams
- Unit weight bucket
- Mechanical sieve shaker with standard sieves, conforming to ASTM E11
- Two flat pans for drying materials

- Slump cone and rod
- Concrete Thermometer
- Normal thermometer
- Measuring tapes
- Cube moulds
- Cylinder moulds
- Two measuring glasses , 1000 ml and 450 ml high
- Trowels, scoops,pails,shovels, wheel barrow
- Compression testing machine
- Drying Oven (convection 110 deg C)
- Water tank for curing
- Applicable ASTM,ACI,PCI literature
- Laboratory mixer

Introduction of Conventional building work.

In this new era of construction, several different building systems orientated for housing appeared in India. Most of these systems were introduced from first world countries and many were not able to fit into the Indian construction standards. On the other hand, many of these systems were functional in countries lacking earthquake trouble and the majority of them employed foreign supplies. In this case of India, structural systems were supposed to be designed according with earthquake standard E-030 of the National Building Code. On the other hand, Indian researchers in the last 20 years developed different solutions for building systems using locally available materials and most of them were tested and had the approval and registration of BMTPC. The present work introduced five construction systems that are approved by BMTPC. The economic issues are also the main reason for these different building systems. So we have to select the most viable and low cost construction for avoid the financial problems.

The major bulk of building construction work is small renovation, such as addition of a room, or renovation of a restroom. Often, the owner of the property work as the introspect, paymaster, and design team for the entire project. Although, all building construction projects consist of some elements in general – design, financial, estimation and legal considerations. Many projects of different sizes reach unwanted consequences, such as structural failure, cost overruns, and/or lawsuits. Due to such reasons, those with expertise in the field make comprehensive plans and keep cautious supervision throughout the project to ensure a positive result.

Commercial building construction is procured privately or publicly utilizing various delivery methodologies, including cost estimating, hard bid, negotiated cost, conventional, organization contracting, construction management-at-risk, design & build and design-build bridging.

Residential construction practices, technologies, and resources should match to the regulations of local building authority and codes of practice. Materials readily accessible in the region usually dictate the building materials used (e.g. brick vs stone vs concrete). Cost of building construction can vary considerably based on site conditions, local policy, economy of scale (custom designed residences are often more costly to build) and the availability of trained workers and labours. As residential building construction (as well as all other types of construction) can produce a lot of waste, so careful planning again is required here.

Aim, Objectives and scope of Precast Concrete Technology.

AIM :

The study aims to evaluate the effective use of Precast Concrete Construction in building systems in order to achieve affordable mass housing and also to minimize the intensive use of building materials.

OBJECTIVE AND SCOPE FOR THE STUDY :

1. To analyse and to study of Precast Concrete in construction management aspect in terms of cost and time.
2. To analyse the cost of structure due to savings in materials.

3. To familiarize with the Precast Concrete construction methodology with due focus on the difference from the conventional construction.
4. To Identify and understand, pros and cons of Precast Concrete in residential construction (mass housing).

Limitation and Methodology for the Study.

The study will focus only in construction phase cost, time and quality of precast concrete construction vs conventional construction.

METHODOLOGY :

The research methodology is divided in two parts; theoretical aspect discusses the building design using Precast construction technology according to design manual, installation manual, and system requirements. Practical aspect concerning cost of operation process, In addition to that, data was gathered from different projects using Precast construction system concerning construction cost & installation procedures on site and is compared against the conventional building. The cost, time, quality and quantity analysis is done for Precast construction and the comparison of conventional building is done to known the feasibility of the Precast construction technology and the conventional system.

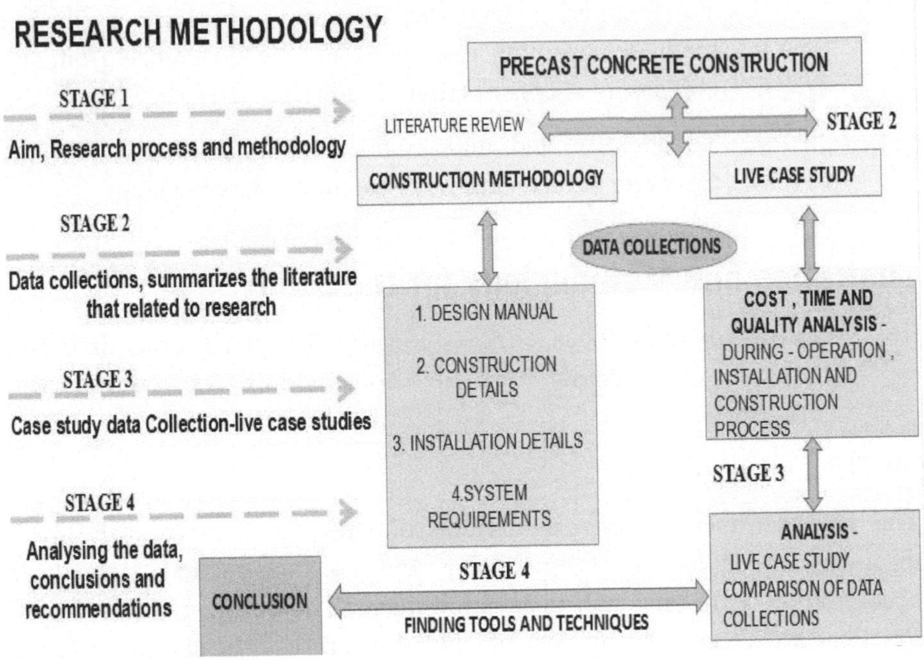

Fig – 36 - Research Methodology

Precast Construction Yard.

All activities related to the construction of reinforced concrete precast segments will be carried out in the precast yard, covering approximately 40,000 square metres of area, including: 1. Rebar storage & rebar jigs area (rebar cage fabrication area). 2. Segment casting area (moulds), 3. Stores & workshop area, 4. Segment storage area, 5. Site offices & basic facilities for staffs.

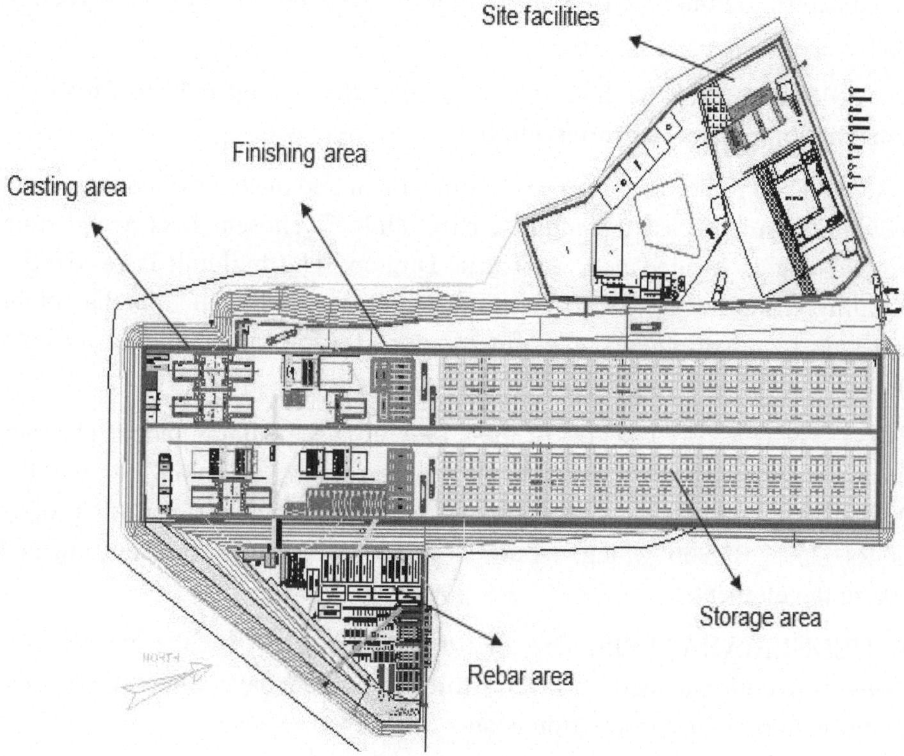

Fig - 37 - Precast Construction Yard

A Comparative Study On Precast Construction And Convetional Construction.

A Comparative Study on Precast Construction and Conventional Construction. International Research Journal of Engineering and Technology (IRJET) by P.Karthigai Priya1,M.Neamitha2 (Aug 2018).

The main objective of this project is to find out an optimal construction technique which has a minimal cost and gives a minimum duration of time by doing a comparative study between precast and conventional construction.

1. To compare the cost and time of precast construction vs conventional construction.

2. To determine the type of construction that would reduce the construction cost.

3. To determine the method of construction that would help to finish the construction project in shorter duration.

CASE STUDY: For the purpose of comparison and analysis of conventional technique and precast technique a case study is chosen. It is a real time project that is located near Ambattur Dunlop. The building is residential building comprising of stilt plus nineteen floors and each floor has eight flats. Cost estimation and scheduling is done for the same building for both precast construction and conventional construction.

COST ANALYSIS: The cost of the entire project including labour charges, cost of material in conventional construction is found to be Rs. 24,79,83,764. Whereas in precast construction the total cost for this project is Rs. 25,17,01,966 which includes the cost of casting, transporting and fixing the elements.

SCHEDULE ANALYSIS: The total duration required for the construction using conventional method of construction is 705 days whereas the total duration in precast construction is only 598 days.

CHECK LIST: Checks are to be done mainly while erecting, transporting and manufacturing. So data has been collected regarding various checks need to be done in precast construction and a check list is prepared.

RISK ANALYSIS: A questionnaire is done to know about the various risks and their level of impact on and off the people on the site. It is prepared based on the factors or conditions that cause injury such as due to heavy lifting, structural collapse, improper barricades and braces, working with heavy equipment etc. The questionnaire was surveyed from various people working in the field of precast.

INFERENCE: After comparison it is found that the cost of precast construction is 1.4% higher than that of the conventional construction. The precast construction can be completed at a rate of 15.17% earlier than the conventional construction. A questionnaire is prepared and analysed for the various risks on precast construction. Charts are drawn to know the impact of injury on various factors in precast construction and they are analysed and the results are provided. Earlier a questionnaire is done to know the importance of precast, so a comparative survey study on conventional and precast is done based on several factors.

Quantitative Analysis Of Precast And Cast In-Situ Residential High Rise Building.

Quantitative Analysis of Precast and Cast In-situ Residential High Rise Building International Research Journal of Engineering and Technology (IRJET) by Akshay Jagannath Rajagopal (Jun 2020)

The main objective of this paper is to find out which construction technique is suitable for high rise buildings. This is done by finding out the cost and time for completion of the project. To find out the appropriate method for building of high rise structures, we "Compare the cost and time for both precast and cast in-situ building".

METHODOLOGY - A plan of dimension 1380 sqft is drawn using Auto Cad, the carpet area is 1198 sqft. Also, a 3D view of both precast and conventional building are drawn to show the difference between precast and conventional construction and to show the G+7 view of the building which consists of 48 houses (8 houses per floor).

	Precast Construction	Cast in-situ	Difference in cost	%
Total Cost	18,31,28,275	17,0349,135	1,27,79,140	7.5
Cost Per House	38,15,173	35,48,940	2,66,233	7.5
Cost Per SqFt	2,765	2,572	193	7.5

Comparative Cost Analysis

Scheduling is done for both precast and conventional construction using Primavera P6. The time for completion of project when it is done using precast method is 244 days while for conventional is 544 days. The estimation of quantities is done except for foundation. The slab cycle for precast building is 11 days. The time for completing one floor using conventional method is 30 days.

The quantity estimation is done based on the following –

i) The grade of concrete is M20 and the mix design for both precast and conventional is 1:1.5:3,

(ii) The cement mortar is in the ratio 1:6, (iii) For precast construction, the quantity is estimated based on the volume of the panels, (iv) The quantity of materials required for foundation is not calculated, instead a sum of Rs.10,00,000 is allotted in the rate analysis, (v) The floor to floor height is 3 m, (vi) This dimension of brick is taken as 190mm x 90mm x 90mm and, (vii) The cost of water for precast and cast in-situ are Rs. 50,00,000 and Rs.1,00,00,000, this is due to the large quantity of water needed for cast in-situ. The volume of one house is 31.5 m^3. For 48 houses, the total volume is 1512 m^3. As per the above regulations, the quantities of major components are as follows- (I) For Precast- Cement- 12096 Bags, Sand- 14742 cft, Steel- 12 tons, Aggregates- 31095 cft and GP2 Grout- 8064 bags. (II) For Conventional- Cement- 9072 Bags, Sand- 62855 cft, Steel- 30 tons, Aggregates- 10000 cft and Brick- 9,82,800 nos. The quantity of paint is found out as 1,230 liters which is split into 2 (for outer and inner). Tractor Emulsion is used and for inner Semi-Gloss is used.

The rate analysis is done based on the market rate in Bengaluru, India. The cost of materials which are give above are as follows- Aggregates- ₹3,900 per cubic foot, Cement- ₹280 per bag, Gp2 grout- ₹450 per bag, Sand-₹960 per cubic foot, Steel-₹42,000 per ton, Brick-₹8 per number, Tractor Emlsion-₹350 per liter and Semi-Gloss-₹635 per liter. The labour charge is estimated as ₹600 per person per day. The cost for setting up the precast plant is estimated as ₹50,00,000. The cost estimated for equipment for both precast and cast in-situ are ₹35,00,000 and ₹10,00,000 respectively. A sum of ₹5,00,000 is added as miscellaneous amount. The cost of foundation, water, plumbing and sanitation, electrical, glass and window panes, tiles, doors and windows for precast is estimated to be ₹1,96,00,000 and for cast in-situ is ₹2,46,00,000. The rate of constructing precast building is ₹18,31,28,275 and for conventional is ₹17,03,49135. Time required for completing pre-cast is 244 days and for conventional is 544 days.

INFERENCE:

Although, the cost for completion of Conventional Building is Less than 7.5% less than Pre-Cast Building, the days to complete exceeds by 300 days. The major outcome of this study is, although the initial amount of precast construction is high, it has advantage over cast in-situ as time consumed is less, which is a huge beneficiary.

Analysis and Results of Precast and Conventional methods of construction for Green Building.

1. INTRODUCTION:

The well-known real estate developer Sobha has brought an amazing and affordable residential project named Dream Acres at a really feasible location in Bangalore. The vast project will be spread over an area of 81 acres making it a highly spacious project in the area. The project will consist of 2 BHK and 2 BHK (Large) Mid - Segment Houses.

The Sobha Dream Acres Apartments at Bangalore is located in the closest proximity to the IT and ITES companies, colleges, malls and schools. The project is known to be the best in the area for its 80% open space availability, precast technology and other high- end construction techniques used in the construction.

2. Precast Type - The Dream Acres, Bangalore

Type: Township

Area: 81 acres

No. of units: 6500 : 1 BHK - 59.96 Sq. mts to 61.43 Sq. Mts, 2 BHK Regular - 69.28 Sq. mts to 94.02 Sq. Mts and 2 BHK Large (phase -2) - 87.00 Sq. mts to 112.43 Sq. Mts.

Type Of Apartment	Mid Segment Houses
Project Stage	Ongoing / Under Construction
Land Area	81 Acres
Type Of Flats	1 BHK, 2 BHK, 2 BHK (Large)
No of Units	6500+ units
Open Space	80%
SBA	650 Sq Ft – 1210 Sq Ft
Project Location	Panathur Main Road, Off ORR, Balagere, Bangalore - 560087
Project Zone	East Bangalore
Total Towers	10 Blocks (Phase 1)
Total Floors	G+13 & 14 Floors
Approvals	BDA/BBMP
Rate Per Sq Ft	Rs. 5150/-
Possession	2018
Budget	Rs. 40 Lakhs Onwards

The Dream Acres Project

3. Project Management

Monitoring: Assigning site engineers to check each activity on site.

Recording: Compiling the monitored data's ie based on DPR(Daily Progress report)

Tracking: Ensuring the work is done based on Master construction plan.

Reporting: Report all the data's to client on weekly basis.

Quality check: Ensuring the quality with quality checklist for each activity.

Quality surveying: Checking the material procurement bill produced by HNS contractor. Done based on material indent generated by vendor / contractor. Running account bill submitted by contractor will be checked and passed within 30 days from the date of submission.

Meetings: Organizing and coordinating a common meeting with client and contractors of many disciplines on weekly /monthly basis. Project manager creates Agenda before meeting and MOM(Minutes of meeting) after meeting.

Scope of services: Client Total Representation, Program Planning, Design Team Management, Time Scheduling, Project Documents Review, On-site Contractors Co- ordination, Quality Control, Health & Safety Provisions Co-ordination and Ensuring all necessary Compliance.

Cost Management: Quantity Survey, Budgeting, Tendering Process, Contract Administration, Value Engineering, Cash-flow Management, Bill Certification and Cost Control.

4. <u>The Dream Acres - Building Data</u>

    1. Column size (Base to 3rd) 300 mm × 675 mm Column size (4th to 9th) 230 mm × 675 mm Column size (10th to 13th) 230 mm × 600 mm Column size (14th to 18th) 230 mm × 525 mm .
    2. Beam size = 230 mm × 450 mm,
    3. Thickness of slab = 115 mm,
    4. Hollow core slabs - 150, 200, and 265 mm,
    5. Stair case slab thickness = 150,
    6. Parapet wall 1 m Height = 230 mm,
    7. Live load = 2 kN/m2,
    8. Floor Finish = 1 kN/m2,
    9. 9. Yield strength of steel = 415 MPa,
    10. Density of concrete = 25 kN/m3
    11. 5% Damping.

Overall built-up Area (Precast+CIS) : 10 Million Sq. ft.

Built-up area planned in Precast : 3.79 million Sq.ft.

Built-up area of A series (A1-A7) : 8.82 lacks Sq. ft.

Total apartments : 1500 apartments; 12 apartments per floor

Structural system (Sub-structure) : Raft foundation (CIS)

Structural system (Super-structure) : Hybrid Precast Construction (Shear walls)

Basement, Ground & terrace work : Cast In Situ 1st to 18th floor

Precast beams, wall panels, balconies, claddings, staircase, & hollow core slabs

Total precast concrete volume (m3) = 16000

Total Nos. of Precast elements (Tilts + HCS) = 25000

No. of units: 6500

- 1 BHK - 59.96 Sq. mts to 61.43 Sq. mts
- 2 BHK Regular - 69.28 Sq. mts to 94.02 Sq. mts
- 2 BHK Large (phase -2) - 87.00 Sq. mts to 112.43 Sq

5. The Dream Acres - Site Works

The sequence of erection shall be as follows:

1. Installation of precast wall panels above cast-in-situ slab.
2. Provide temporary props/jacks for restraining of the walls laterally.
3. Grout the connection between the wall panels & ground floor slab and the joint between each wall panel.
4. Installation of precast slab panels above the erected precast wall panels.
5. Screed concrete above the slab after placing of electrical conduits/fittings.
6. Installation of the wall panels over the floor slab.
7. Installation of the roof panels such as parapets etc.

6. The Sobha Dream Acres - Stages Involved in Construction

7. The Sobha Dream Acres - Design and Implementation:

Hybrid concrete construction is a method of construction which integrates precast concrete and cast in-situ concrete to make best advantage of their different inherent qualities. The accuracy, speed and high-quality finish of precast components can be combined with the economy and flexibility of cast in-situ concrete. In this hybrid construction foundation,

Substructure, terrace work & all the shear walls for all floors were completed using conventional CIS method. Superstructure consisted of precast construction of beams, balcony, staircase, wall panels & hollow core slabs. The ambitious real estate program features 13 towers of 18 floors built in a record time by precast construction methods supported by a purpose-built factory powered by Vollert equipment. After casting on the production line, the panel is brought to its final location for straightening and bracing installation. Locally sourced non-shrink grout is inserted into the cavity by injection. After injection, the bracing around the precast panel may then be removed after only 3 days.

8. The Dream Acres - Connections Details:

The vertical members (shear walls) were constructed in CIS considering the economy and design aspects along with erection plan of other members. This hybrid design facilitated the overall faster erection and allowed for timely production & delivery of other precast members. The connection were simple and special dowel bars, corbels and provisions for connection were planned in advance for the members as per the design. The joints were made monolithic by grouting them with non-shrink grout. Onsite, wet grouted joints for horizontal connections between wall to wall, slabs to wall, termed as stitched joints were selected.

9. <u>The Dream Acres - Challenges Experienced:</u>

As the project was located in seismic zone IV, the suitable design was adopted. Lack of skilled manpower with the knowledge of precast industry, operation of various machineries & erection, which was overcome by providing detailed training. Although, European and other standards were available for ready reference but unavailability of IS codes was deemed important. The designers from India had only few Indian Standard to refer.

10. <u>The Dream Acres - Result Achieved:</u>

6 buildings of two buildings G+18 floors were completed in 14 months of timeline with superior quality was made possible through precast technology. Achieved a slab cycle of 10-12 days for a slab area of 7000 Sq. ft. Cost & time optimisation with the use of hollow-core slabs. Reduction in concrete & steel factor per Sq. ft. of built-up area. Automation & mechanisation of construction project improved overall productivity. Improved site safety & considerable reduction in wastage, dust & noise on site, thus reduced ecological footprint of project.

11. <u>The Dream Acres – Conclusion:</u>

Hybrid precast instead of usually considered 'all precast or no precast' approach offers more benefits in terms of time, cost, and quality. Repetition of precast components is essential in order to meet quantity for cost effectiveness. Improved quality control, reduction of construction time, construction waste, dust & noise on site, and labour requirement on site. it results in higher useful gross floor area which contributed to significant cost benefits. A considerable reduction in steel factor can be achieved even for seismic zones IV & V. Precast technology allowed for flexible design and longer clear spans in the non-tower areas like parking. Precast is smart way to achieve the sustainability objectives of Green Building.

12. The Sobha Dream Acres - Vollert equipment:

The ambitious real estate program features 13 towers of 18 floors built in a record time by precast construction methods supported by a purpose-built factory powered by Vollert equipment.

Merits And Demerits Of Precast And Conventional Methods.

The following tables explain about the : Complete details about the Precast technology and the conventional systems. Detailed estimation / quantity for Precast Technology and Conventional system is done. Cost and time comparison is done to known the feasibility.

1. Comparison of Precast system and conventional systems:

PARTICULARS	DREAM ACRES (PRECAST)	SHOBHA WINCHESTER (CONVENTIONAL)
Construction speed	Very rapid speed of erection. Rapid construction on site - 462 DAYS.	Comparatively slow construction On site casting, so reinforcement laying & fixing, formwork. - 605 DAYS.
Quality control	Good quality control. Quality is checked under factory conditions periodically. Errors are avoided as they are computerized high durability is achieved.Curing also done in factory conditions.	Quality may affect due to site conditions, due to bad supervision, unskilled labour and procurement of materials. QC is checked regularly for achieving good quality.
Environmental conditions	Weather is eliminated as a factor-you can cast in any weather and get the same results.	Environmental conditions like temperature, humidity can affect on performance of concrete.
Labour Requirement	Fully automated in production stage with help of unskilled labours, but for erection technically skilled labours is needed.	More Labours required on site. Mix of skilled and unskilled labours are present according to the works.
Manufacturing conditions	High quality is achieved because of the controlled conditions in the factory. And they are of higher quality.	RCC is to casted on site & the site conditions are not regularized, so it may affect on strength.

PARTICULARS	DREAM ACRES (PRECAST)	SHOBHA WINCHESTER (CONVENTIONAL)
Cost	Rs.1600/sq.ft	Rs.2000/sq.ft
Connections	Connections are quite difficult 50 mm gap is maintained.	Connections are not needed.
Size & Shape	Repeated precast elements, so few moulds are prepared priorly with all the dimension check.	In-situ concreting is suitable where the building is in uneven shape & there are no repetitive shapes because of human errors.
Size Limitation	No limitation for size.	Because panel size is limited, precast concrete cannot be used for two-way structural systems.
Cost	It is comparatively economical when the building having no regular shapes.	Economics of scale demand regularly shaped buildings
Finishing	The finishing are perfect as it s all computer aided deigns	Manual error occurs always so finishing is not that much perfect.

PARTICULARS	DREAM ACRES (PRECAST)	SHOBHA WINCHESTER (CONVENTIONAL)
Dismantling	Dismantling can be so much easier as they can be removed by disconnecting all the joints.	Dismantling is not easy ,machinery or much labour s needed.
Carbon emission	5% less carbon emission.	Its more comparing to precast.
Services	Electrical conduits are fixed in wall panels and hollow core slabs in factory itself Plumbing pipelines are inlayed and then highly qualified water sealants are used thus making the building waterproof.	Electrical conduits are laid at the time of roof slab construction. Plumbing pipelines are designed and waterproof of the building depends on workmanship.

2. Detailed estimation / quantity for Precast System- Live case study

Detailed estimation / quantity for Precast System

PRECAST METHOD : COLUMN (PER FLOOR)	
Precast column	**M25 Grade**
Column Size	**0.35x0.35x2.820 M**
Total volume	0.35 m3
Steel required	46 kg
R.M.C rate for 1m3 concrete for 0.35 m3 concrete cost.	Rs.5531 = (5531X0.35) = Rs.1936
Labor cost	Rs.1500
Erection cost	Rs.2500
Total cost required for 1 column Concrete	1936
Steel cost	= 46X53.9(Rate/KG) = 2479
Labour cost for precast production	Rs.1500
Total column cost	Rs.9915
Total no of column required for 1 floor 36 No's. **Total cost = 9915X36 = Rs.356940** **1m3 Rate for column = Rs.28329.**	

PRECAST METHOD : BEAM (PER FLOOR)	
Precast Beam	**M25 Grade**
Total volume of Beam	14.202 m3
Steel required	2011.18 kg
R.M.C rate for 1m3 concrete	Rs.5531
For 14.202 m3 concrete cost	Rs.78551
Labor cost for 57 beam	500x57 = Rs.28500
Erection cost for 1 beam	RS.2500
Total 57 no's of Beam	2500x 57= Rs.142500
Total cost required for 57 no's of beam	
Concrete	Rs.78551
Steel cost	Rs.108403
Labour cost	Rs.28500
Erection cost	Rs.142500
Total beam cost = Rs.357954 **1m3 Rate for Beam = Rs.25205**	

PRECAST METHOD : LINTEL (PER FLOOR)	
Total volume of lintels	0.94 m3
Steel required	72.16 kg
R.M.C rate for 1m3 concrete	Rs.5531
For 0.94 m3 concrete cost	Rs.5199
Labour cost	Rs.25
Erection cost for 1 lintel	Rs.40
Total cost required for 48 no's of lintel	
Concrete	Rs.5199
Steel cost	Rs.3889
Labour cost	Rs.1500
Erection cost @ site	Rs.4000
Production cost	Rs.5000
Total lintels cost = Rs.19588 **1m3 Rate for lintels = Rs.434**	

PRECAST METHOD : SLAB (PER FLOOR)	
Total cost Required for 92 panels	Rs.200640
Site Erection cost-Total 92 no's of panels	= 100x92nos = Rs.92000
Total cost required for 92 no's of panels	
Total slab panel cost	Rs.292640
Total volume Screeding of 40 mm.	8.32 m3
Production & post-production cost	75000
For 1 m3 m 25 grade rate of concrete	Rs.5531
Steel required	750 kg - 750x53.9 = Rs.40425
Total Screeding cost	
Concrete cost	8.32x5531 = Rs.46018
Steel cost	175x53.9 = Rs.9434
Total cost for screeding = Rs.55452 **Total cost for slab = Rs.497474**	

PRECAST METHOD : LINTEL WITH CHAJJA (PER FLOOR)	
volume of 1 lintels with chajja	0.78 m3
Steel required	188.19 kg
R.M.C rate for 1m3 concrete	Rs.5531
For 0.78 m3 concrete cost	Rs.4314
Labour cost	Rs.125
Erection cost for 1 lintel with chajja	Rs.150
Total 28 no's of lintels with chajja	Rs.4200
Production cost	Rs.20000
Concrete	0.78x5531 = RS.4314 X 28 = 120792
Steel cost	188.19x53.9 = 10143
Labour cost (@site)	Rs.2000
Erection cost (@site)	Rs.4500
Total lintel with chajja cost = Rs.161635 **1m3 Rate for lintel with chajja = Rs.7525**	

PRECAST METHOD : STAIRCASE (PER FLOOR)	
Total volume of staircase	0.48 m3
Steel required	36 kg
R.M.C rate for 1m3 concrete	Rs.5531
Labour cost	Rs.125
Erection cost for 1 staircase	Rs.1000
Total 4 no's of staircase	4x1000 = Rs.4000
Total cost required for staircase	
Concrete	0.48x5531= Rs.2655 x 4 = 10620
Steel cost	36x53.9 = Rs.1940 x 4 = 7760
Labour cost	Rs.5000
Production cost	Rs.12000
Total no of Staircase required for 1 floor 4 No's Total staircase cost = Rs.40380 **Total cost per staircase = Rs.10095**	

3. Detailed estimation / quantity for Conventional System - Live case study

CONVENTIONAL METHOD : COLUMN (PER FLOOR)	
Column Size	**0.35x0.35x2.820 M**
Total volume	0.35 m3
Steel required	46 kg
Conventional rate for 1m3 concrete for 0.35 m3 concrete cost	= RS.8188 = 8188X0.35 = RS.2866
Labor cost	Rs.5000
Total cost required for 1 column Concrete	RS.2866
Steel cost	= 46x53.9 = RS.2479
Shuttering Rs.25/sqft	= Rs.1000x36=36000
Total column cost	= RS.6345
Total no of column required for 1 floor 36 No's. **Total cost = 6345x36 = Rs. 269420** **1m3 Rate for column = Rs.21383**	

CONVENTIONAL METHOD : BEAM (PER FLOOR)	
Total volume of Beam	14.202 m3
Steel required	2011.18 kg
Conventional rate for 1m3 concrete	Rs.8237
For 14.202 m3 concrete cost	Rs.116982
Total cost required for 57 no's of Beam	
Shuttering Rs.25/sqft	= Rs.1000x57=57000
Labor cost	= Rs.5000
Concrete	= 14.202x8237 = Rs. 116982
Steel cost	= 2011.18x53.9 = Rs.108403
Total Beam cost = Rs.287385 **1m3 Rate for Beam = Rs.20236**	

CONVENTIONAL METHOD : LINTEL (PER FLOOR)	
Total volume of lintels	0.94 m3
Steel required	72.16 kg
Conventional rate for 1m3 concrete	Rs.8237
For 0.94 m3 concrete cost	Rs.7743
Shuttering Rs.25/sqft	= Rs.1000
Labour cost	= Rs. 1000
Concrete	Rs.7743
Steel cost	Rs.3889
Total lintels cost = Rs.13632 **1m3 Rate for lintels = Rs.302**	

CONVENTIONAL METHOD : LINTEL (PER FLOOR)	
Total volume of lintels	0.94 m3
Steel required	72.16 kg
Conventional rate for 1m3 concrete	Rs.8237
For 0.94 m3 concrete cost	Rs.7743
Shuttering Rs.25/sqft	= Rs.1000
Labour cost	= Rs. 1000
Concrete	Rs.7743
Steel cost	Rs.3889
Total lintels cost = Rs.13632 **1m3 Rate for lintels = Rs.302**	

CONVENTIONAL METHOD : SLAB (PER FLOOR)	
Total volume of Slab	23.44 m3
Steel required	2760 kg
Conventional rate for 1m3 concrete(RMC) For 23.44 m3.	Rs.8843 x23.44m3 = 207280
Shuttering Rs.25/sqft	= Rs.10000
Labour cost	= Rs.8000
Total cost required for Slab Concrete	Rs.207280
Steel cost = 2760x53.9= Rs.148764 **Total Slab cost = Rs.374044**	

CONVENTIONAL METHOD : LINTEL WITH CHAJJA (PER FLOOR)	
volume of 1 lintels with chajja	0.78 m3
Steel required	188.19 kg
Conventional rate for 1m3 concrete	Rs. 5531
For 0.78 m3 concrete cost	Rs.4314
Total cost required for 28 no's of lintel with chajja	
Concrete	28x4314=RS.120792
Steel cost	188.19x53.9=RS.10143
Labour Cost	1500
Shuttering Cost	1500
Total lintels cost = Rs.133935 **1m3 Rate for lintels = RS. 6235**	

CONVENTIONAL METHOD : STAIRCASE (PER FLOOR)	
Total volume of staircase	0.48 m3
Steel required	36 kg
Conventional rate for 1m3 concrete	Rs.5531
? Total cost required for stair case	
Concrete	0.48x5531 = Rs.2655 x 4 = 10620
Steel cost	36x53.9 = Rs.1940 x 4 =7760
Labour cost	1000 x 4 = 4000
Shuttering Cost	2000 x 4 = 8000
Total staircase cost	30380
Total no of Staircase required for 1 floor 4 No's **Total cost per staircase = 30380 / 4 = RS.7595**	

4. Cost comparison between Precast and Conventional system.

SI.NO	COMPONENTS OF THE BUILDINGS	PRECAST CONSTRUCTION	CONVENTIONAL CONSTRUCTION	COST DIFFERENCES
1.	**Column**	Rs.356940	Rs.269420	Rs.87520
2.	**Beam**	Rs.357954	Rs.287385	Rs.70569
3.	**Lintel**	Rs.19588	Rs.13632	Rs.5956
4.	**Slab**	Rs.497474	Rs.374044	Rs.123430
5.	**Lintel with chajja**	Rs.161635	Rs.133935	Rs.27700
6.	**Staircase**	Rs.40380	Rs.30380	Rs.10000
	Total cost	**Rs.1453332**	**Rs.1108796**	**Rs.325175**

5. Time comparison between Precast and Conventional system.

PRECAST BUILDING CONSTRUCTION

SI.NO	DESCRIPTIONS - DREAM ACRES, BANGLORE	DURATIONS
1.	**SUB STRUCTURE** - (Site cleaning, Earthwork, Foundation, Basement, Soil filling & Consolidation)	88 days
2.	**SUPER STRUCTURE** – (Wall panels framing and Roofing slabs)	190 days
3.	**FINISHING WORKS** – (Electrical, Plumbing Painting, Tiling, and Windows, Extra items)	184 days
	TOTAL DURATIONS	**462 days**

CONVENTIONAL BUILDING CONSTRUCTION

SI.NO	DESCRIPTIONS - SHOBHA WINCHESTER	DURATIONS
1.	**SUB STRUCTURE** - (Site cleaning, Earthwork, Foundation, Basement, Soil filling & Consolidation)	88 days
2.	**SUPER STRUCTURE** – (Wall and Roofing slabs)	333 days
3.	**FINISHING WORKS** – (Electrical, Plumbing Painting, Tiling, and Windows, Extra items)	184 days
	TOTAL DURATIONS	**605 days**

6. CONCLUSION

1. A sample analysis for a floor alone (small scale) is taken as a example in which precast structure is costlier than conventional, but Project labor and time are very much less than Conventional. Also, Precast is much economical than Conventional method when the project scale increases. Study says that In a large scale projects with high rise buildings, Precast is 20-25% economical than Conventional.

2. From the analysis of the case studies conducted, it was found that the cost of construction of Precast structures was 14.5cr. whereas in case of Conventional building it was 17.1cr.

3. Based on the above analysis, we can infer that a construction of Precast structure is quicker than a Conventional structure. Hence we can infer that Precast Structures requires very less labor & Time saving building solution.

4. As of now, Precast method is the best solution for affordable mass housing in India. To conclude that, Precast method of construction is an alternative and efficient solution for building construction techniques.

5. Therefore, this is the good solution for fast delivery of mass dwelling / housing and plays a vital role in reducing huge urban housing shortage in India.

Case Study on use of Precast Concrete Technology for Construction of Barwa housing Project in Doha, Qatar.

A real-time case study The Barwa City Project, Phase 1, is a residential Project (102 504 m2 GFA, 1 Billion Euro project) consisting of 128 apartment buildings in different sizes. The houses are classified in three different types Type A and B with up 7 floors (G+6) and type C with 2 floors (G+1).To keep the construction time short and ensure a uniform high quality standard, precast wall and slab elements have been chosen as the most favorable structural system for the buildings, Apart from buildings construction outside installation of all utilities, the road works and the landscaping are also scope of work.

Type A (G+6) = 12 buildings

Type A (G+5) = 16 Buildings

Type A (G+4)= 16 Buildings

Type A (G+3) = 4 Buildings

Type B (G+6) = 12 buildings

Type B (G+5) = 16 Buildings

Type B (G+4)= 16 Buildings

Type B (G+3) = 4 Buildings

Type C(G+1) = 32 Buildings

Total number of Buildings are Type A =12+16+16+4=48 Apartment buildings,

Type B = 12+16+16+4 = 48 Apartment Buildings,

Type C = 32 Buildings

Total number of Apartment buildings are = 48+48+32=128 Apartment buildings.

Remarks:

(1) I need total number of elements used for 128 Apartment buildings data (Precast elements + Hollowcore slabs) including both Precast and Pre-stressed how many Total elements are used.

(2) Total Volume of concrete used for construction of 128 Buildings.

(3) 128 Apartment buildings for 6000 residential units. Please let me know in Type A each floor is how many flats and each flats has how many units. Then i will get answer of Type A all data as well as type B all data of residential units.

(4) Type C was stereo Type Apartment buildings so let it clear Type C.

For all panels slump range is 120 mm and tolerance limit is + 30 mm and – 30 mm.

How to Prepare to face a job interview in precast concrete Technology?

Question # 1

Your coworker highlights your mistakes in front of everyone, how do you handle the situation?

Answer:-

Admit to the mistake without being emotional, but then discuss how you are being proactive in getting it fixed. Lastly, pull the co-worker aside later on to tell them that you'd appreciate it if they gave you the feedback 1:1 first before throwing you under the bus.

Question # 2

What critical component of this position Regarding Erector Precast Concrete makes the work challenging?

Answer:-

Heading information: This should include job title, pay grade or range, reporting relationship (by position, not individual), hours or shifts, and the likelihood of overtime or weekend work.

Summary objective of the job: List the general responsibilities and descriptions of key tasks and their purpose, relationships with customers, coworkers, and others, and the results expected of incumbent employees.

Qualifications: State the education, experience, training, and technical skills necessary for entry into this job.

Special demands: This should include any extraordinary conditions applicable to the job Regarding Erector Precast Concrete (for example, heavy lifting, exposure to temperature extremes, prolonged standing, or travel).

Job duties and responsibilities: Only two features of job responsibility are important: identifying tasks that comprise about 90 to 95 percent of the work done and listing tasks in order of the time consumed (or, sometimes, in order of importance).

Question # 3

Tell me about the most fun you have had on the job?

Answer:-

When answering this question, discuss situations where you completed tasks benefitting your previous employers.

Question # 4

How do you think you might fit this position Regarding Erector Precast Concrete?

Answer:-

An important part of research before the interview is what the company does and how the job role relates to that. This includes the company philosophy and working methods. Questions such as this seek to find out how a candidate will fit into the organisation Regarding Erector Precast Concrete. Answer positively; including practical examples of how you anticipate you would perform in the new role.

Question # 5

You have a project due in one hour but a more important emergency that affects business needs to be fixed immediately, what do you do?

Answer:-

Focus on the issue that impacts the business most first.

Question # 6

Can you tell me a little about yourself?

Answer:-

This question seems simple, so many people fail to prepare for it, but it's crucial. Here's the deal: Don't give your complete employment (or personal) history Regarding Erector Precast Concrete. Instead give a pitch-one that's concise and compelling and that shows exactly why you're the right fit for the job. Start off with the 2-3 specific accomplishments or experiences that you most want the interviewer to know about, then wrap up talking about how that prior experience has positioned you for this specific role.

Question # 7

Tell me about a difficult decision you've made in the last year Regarding Erector Precast Concrete?

Answer:-

We all have difficult decisions in our lives. Show how you were able to arrive at it and then how you decisively acted.

Question # 8

Tell me why do you want this job Regarding Erector Precast Concrete?

Answer:-

Bad Answer: No solid answer, answers that don't align with what the job actually offers, or uninspired answers that show your position is just another of the many jobs they're applying for.

Good answer: The candidate has clear reasons for wanting the job that show enthusiasm for the work and the position, and knowledge about the company and job.

Question # 9

What motivates you at the work place?

Answer:-

Keep your answer simple, direct and positive. Some good answers may be the ability to achieve, recognition or challenging assignments.

Question # 10

Can you explain why you changed career paths Regarding Erector Precast Concrete?

Answer:-

Don't be thrown off by this question-just take a deep breath and explain to the hiring manager why you've made the career decisions Regarding Erector Precast Concrete you have. More importantly, give a few examples of how your past experience is transferable to the new role. This doesn't have to be a direct connection; in fact, it's often more impressive when a candidate can make seemingly irrelevant experience seem very relevant to the role.

Question # 11

Do you have any blind spots?

Answer:-

This question is often meant to trick candidates since acknowledgment of blind spots would indicate they were aware of them. Also, do not disclose bad habits or other personal concerns. Let the interviewer find out about your personal flaws through the course of the interview without directly stating these flaws.

Question # 12

Are you willing to work overtime or odd hours?

Answer:-

Be completely honest. You don't want to lie to get the job if you're not going to work the hours required.

Question # 13

Tell me about a time you had to fire a friend?

Answer:-

Hopefully you've never had to do this, but if you did, talk about how hard it was personally to fire anyone but that you did it objectively.

Question # 14

Tell me about a problem that you've solved in a unique or unusual way. What was the outcome? Were you happy or satisfied with it?

Answer:-

In this question the interviewer is basically looking for a real life example of how you used creativity to solve a problem.

Question # 15

What's your dream job?

Answer:-

Along similar lines, the interviewer wants to uncover whether this position Regarding Erector Precast Concrete is really in line with your ultimate career goals. While "an GGL star" might get you a few laughs, a better bet is to talk about your goals and ambitions-and why this job will get you closer to them.

Question # 16

What can you offer us that someone else can not?

Answer:-

Bad Answer: Going negative - if the candidate starts trash talking other candidates, it's a sure sign of a bad attitude. Also, if they can't provide a solid answer, it may show that they lack thorough knowledge of the skills the job requires, and an understanding of where they fit in.

Good answer: The candidate can name specific skills, abilities or understandings they have that apply directly to the job that other candidates are unlikely to have, or are in short supply.

Question # 17

What have you done to improve your skills over the past year Regarding Erector Precast Concrete?

Answer:-

You'll want to be prepare with some very specific examples of what you've done over the last year and what you're currently doing to improve your professional knowledge and skill set as well as anything else you're doing the shows self improvement.

Question # 18

Are you able to relocate if required?

Answer:-

Be completely honest and thoughtful with this one. You don't want to wake up one to find out that you're moving to a new city or state and it may be a major factor in your eligibility for employment. But again, if you don't want to move then the job probably isn't for you.

Question # 19

If you felt like you were hitting the proverbial "wall" and getting burned out, what would you do to re-energize yourself?

Answer:-

Take a break to rest. Work in smaller increments of time to increase focus with breaks in between. Delegate tasks to those that are willing to help.

Question # 20

You are not given the tools you need to be successful. How would you change that Regarding Erector Precast Concrete?

Answer:-

State a business case to your manager / leader as to why you need the tools and make the request for them.

Question # 21

What is the most irritating thing you've experienced about your co-workers?

Answer:-

This question is designed to find out if you get along well on team, with other and whether or not you'll be a fit with the interviewer's organization. It's a trap. Think real hard but fail to come up anything that irritated you about your co-workers. A short positive response is best.

Question # 22

If you have seven white socks and nine black socks in a drawer, how many socks do you have to pull out blindly in order to ensure that you have a matching pair?

Answer:-

if the first one is one color (say, white), and the second one is the other color (black), then the third one, no matter what the color, will make a matching pair. (Sometimes you're not supposed to think that hard.)

Question # 23

Have you ever mentored anyone before? If yes, describe the situation?

Answer:-

Describe a time where you've helped someone else. Mentor ships can be informal so as you've helped someone over a period of time that can certainly count. The key is to highlight how you utilized certain skills/attributes like coaching, teaching, patience, communication skills, and so forth to mentor that person.

Question # 24

What are your thoughts on failure?

Answer:-

Failure happens. It's a part of life. The key is understanding that you can't be perfect at everything and more importantly you're going to learn from failures to come out stronger.

Question # 25

How do you feel about technology at the workplace in general?

Answer:-

It's a great enabler for us to collaborate better as a team, for us to reach customers more efficiently and frequently and I believe it can help any company become more efficient, leaner, and more productive.

Question # 26

What makes a product successful?

Answer:-

Basing on the monetization, these questions give you the chance to prove your personal try. Do not show extremely your optimism and pursue the unreality. Give your answers the reality.

It is useful to predict a five to ten- year- scenario of expectations in order to gain your targets that you set up and it is the period of time to see how your plans and targets are performed.

Therefore, the quality of the product and marketability of the mentioned industry need to be highlighted. This will help you to achieve the interviewer's attention and insurance to you personality and you can get the honest and long- term goals.

Question # 27

What does "collaboration with teammates" mean to you?

Answer:-

Drinking at the water cooler together is not the best example. Think of how you can collaborate with teammates to generate new ideas, to create initiatives to impact the business' success for the better (specifically in the department that you're applying for). For example, if you're applying to marketing, collaboration could mean discussing new ways of social media advertising to reach an audience of over a million people to strengthen the brand awareness of the company.

Question # 28

What did you major in and why?

Answer:-

Tell them your major and the motivations behind why you chose it and how it's helped to prep your of this potential job.

Question # 29

How have you achieved your success?

Answer:-

Discuss stories of how you've progressed over the years to achieve success. People relate best to stories.

Question # 30

If someone had to say something negative to you, what would they say?

Answer:-

Again, be honest about sharing a story here about someone who may not have gotten along with you in the office here and explain how you were able to fix that relationship or change your attitude/action to be a better person / coworker.

Question # 31

What is your biggest regret to date and why?

Answer:-

Describe honestly the regretful action / situation you were in but then discuss how you proactively fixed / improved it and how that helped you to improve as a person/worker.

Question # 32

Describe what a bad work environment would look like to you Regarding Erector Precast Concrete?

Answer:-

There could be a multitude of things to discuss here: Business ethics (wrongdoing), inconsiderate teammates, non-supportive management, a product that does not do what you're promising customers and so forth.

Question # 33

Do you value recognition or pay more? Why?

Answer:-

Either preference is fine, but just remember you have to be able to explain why. If you say recognition, then back that up by describing how achievement really carries weight with you and how you like to feel valued in the work that you do because it validates that you're helping your teammates / customers and so forth. If you choose money, you can also explain that is important to you as validation and you can highlight how money is important to you because of your goals (financial security, providing for your family, and so forth). The key is to be authentic with your answer. However, if you say you value pay more because you're greedy - know that doesn't align usually to most company's values/vision.

Question # 34

Why do you want to join our company?

Answer:-

This is a question that is aimed at finding out whether you know enough about the company and the basic market. The best way to answer this question is to do some research on the company and highlight its positive points.

Question # 35

What classes did you enjoy most in college and why?

Answer:-

Think back to the classes that either resonated with your passion or truly helped you to develop skills that you believe will help you in your career. Talk about those.

Question # 36

How would you estimate the weight of the Chrysler building?

Answer:-

This is a process guesstimate where the interviewer wants to know if you know what to ask. First, you would find out the dimensions of the building (height, weight, depth). This will allow you to determine the volume of the building. Does it taper at the top? (Yes.) Then, you need to estimate the composition of the Chrysler building. Is it mostly steel? Concrete? How much would those components weigh per square inch? Remember the extra step: find out whether you're considering the building totally empty or with office furniture, people, etc. If you're including the contents, you might have to add 20 percent or so to the building's weight.

Question # 37

Suppose there are three light switches outside a room. Inside is a single light bulb, controlled by one of the three switches. You need to determine which switch operates the bulb. You can turn the switches on and off as many times as you wish (they are all off to begin with), but may only enter the room once. There is no one there to help you. The door to the room is closed, and there are no windows, so you cannot see inside. How can you discover which switch operates the bulb?

Answer:-

Do the following steps:

• 1. Turn ON two switches, and leave one OFF.

• 2. Wait a few minutes.

• 3. Turn one switch from ON to OFF. One is now ON and two are OFF

• 4. Enter the room. - If the light is ON, it is controlled by the switch you left ON. - If the light bulb is OFF, touch it. If it is warm it is controlled by the switch you turned ON and OFF. If it is cold, it is controlled by the switch you never turned on.

Question # 38

What other companies are you interviewing with?

Answer:-

Companies ask this for a number of reasons, from wanting to see what the competition is for you to sniffing out whether you're serious about the industry. "Often the best approach is to mention that you are exploring a number of other similar options in the company's industry,". It can be helpful to mention that a common characteristic of all the jobs you are applying to is the opportunity to apply some critical abilities and skills that you possess. For example, you might say 'I am applying for several positions with IT consulting firms where I can analyze client needs and translate them to development teams in order to find solutions to technology problems.'

Question # 39

What have you done to reduce costs, increase revenue, or save time?

Answer:-

Even if your only experience is an internship, you have likely created or streamlined a process that has contributed to the earning potential or efficiency of the practice. Choose at least one suitable example and explain how you got the idea, how you implemented the plan, and the benefits to the practice.

Question # 40

Do you have the ability to articulate a vision and to get others involved to carry it out?

Answer:-

If yes, then share an example of how you've done so at work or college. If not, then discuss how you would do so. Example: "I would first understand the goals of the

staff members and then I would align those to the goals of the project / company. Then I would articulate the vision of that alignment and ask them to participate. From there, we would delegate tasks among the team and then follow up on a date and time to ensure follow through on the tasks. Lastly, we would review the results together."

Question # 41

Why do you want to work in this industry Regarding Erector Precast Concrete?

Answer:-

Make sure you research the industry first. Then find at least 3 core things about that industry that you're passionate about (for example: how their solutions impact clients, their culture, the leadership, etc)

Question # 42

Explain a time when you did not get along with your coworker?

Answer:-

I used to lock heads with a fellows. We disagreed over a lot of things - from the care of civilians to who got what shifts to how to speak with a victim's family. Our personalities just didn't mesh. After three months of arguing, I pulled her aside and asked her to lunch. At lunch, we talked about our differences and why we weren't getting along. It turns out, it was all about communication. We communicated differently and once we knew that, we began to work well together. I really believe that talking a problem through with someone can help solve any issue.

Question # 43

What is your desired salary Regarding Erector Precast Concrete?

Answer:-

Bad Answer: Candidates who are unable to answer the question, or give an answer that is far above market. Shows that they have not done research on the market rate, or have unreasonable expectations.

Good answer: A number or range that falls within the market rate and matches their level of mastery of skills required to do the job.

Question # 44

How do you plan to go by an example for your subordinates?

Answer:-

Sticking to the rules by yourself, working hard and not mind participating on basic tasks is a good answer.

Question # 45

What are three positive things your last boss would say about you?

Answer:-

It's time to pull out your old performance appraisals and boss's quotes. This is a great way to brag about yourself through someone else's words: "My boss has told me that I am the best designer he has ever had. He knows he can rely on me, and he likes my sense of humor."

Question # 46

What do you think we could do better or differently?

Answer:-

This is a common one at startups. Hiring managers want to know that you not only have some background on the company, but that you're able to think critically about it and come to the table with new ideas. So, come with new ideas! What new features would you love to see? How could the company increase conversions? How could customer service be improved?

You don't need to have the company's four-year strategy figured out, but do share your thoughts, and more importantly, show how your interests and expertise would lend themselves to the job.

Question # 47

Explain me about a time when you reached a goal within a tight deadline?

Answer:-

I work well under pressure to meet deadlines without jeopardizing the quality of my work. I have always worked in a fast pace environment where we are constantly under pressure to achieve best results within a time frame.

Question # 48

How do you think I rate as an interviewer?

Answer:-

I think you did fine. I'm sure you've conducted a lot of interviews, and it's probably second nature for you now. Thanks for taking the time to meet with me today. I'm sure you have a lot of things you have to juggle every day.

I'd say you rate at least ten out of ten. The questions you asked seemed spot on. I can tell you guys are working hard to find the perfect applicant for the job. I'm glad I could meet with you.

Question # 49

What features of your previous jobs have you disliked?

Answer:-

It's easy to talk about what you liked about your job in an interview, but you need to be careful when responding to questions about the downsides of your last position. When you're asked at a job interview about what you didn't like about your previous job, try not to be too negative. You don't want the interviewer to think that you'll speak negatively about this job or the company should you eventually decide to move on after they have hired you.

Question # 50

Why do you want to work Regarding Erector Precast Concrete for this organisation?

Answer:-

Being unfamiliar with the organisation will spoil your chances with 75% of interviewers, according to one survey, so take this chance to show you have done your preparation and know the company inside and out. You will now have the chance to demonstrate that you've done your research, so reply mentioning all the positive things you have found out about the organisation and its sector etc. This means you'll have an enjoyable work environment and stability of employment etc - everything that brings out the best in you.

Question # 51

Do you have any question regarding this job Regarding Erector Precast Concrete?

Answer:-

Never ask Salary, perks, leave, place of posting, etc. regarded questions. Try to ask more about the company to show how early you can make a contribution to your organization like. "Sir, with your kind permission I would like to know more about induction and developmental programs?" OR Sir, I would like to have my feedback, so that I can analyze and improve my strengths and rectify my shortcomings.

Question # 52

How do you act when you encounter competition?

Answer:-

This question is designed to see if you can rise the occasion. You want to discuss how you are the type to battle competition strongly and then you need to cite an example if possible of your past work experience where you were able to do so.

Question # 53

Explain an occasion when you had to adapt in the face of a difficult situation?

Answer:-

One of the most useful interview tactics is to remain positive about your work and achievements. This question lets the candidate draw on their own personal history to show how they have been positive and successful in the face of difficulties. Choose a specific occasion to describe, rather than dealing with generic platitudes.

Question # 54

Do you work well under pressure?

Answer:-

Yes.. When it comes down to the wire, the best thing I can to remain focused, have some flexibility, and understand priorities.. Giving them attention in the order they are needed.

Question # 55

How would you rate your communication and interpersonal skills for this job Regarding Erector Precast Concrete?

Answer:-

These are important for support workers. But they differ from the communication skills of a CEO or a desktop support technician. Communication must be adapted to the special ways and needs of the clients. Workers must be able to not only understand and help their clients, but must project empathy and be a warm, humane presence in their lives.

Question # 56

What do you know about the company?

Answer:-

Any candidate can read and regurgitate the company's "About" page. So, when interviewers ask this, they aren't necessarily trying to gauge whether you understand the mission-they want to know whether you care about it. Start with one line that shows you understand the company's goals, using a couple key words and phrases from the website, but then go on to make it personal. Say, "I'm personally drawn to this mission because..." or "I really believe in this approach because..." and share a personal example or two.

Question # 57

How do you keep each member of the team involved and motivated?

Answer:-

Many managers mistakenly think that money is the prime motivator for their employees. However, according to surveys by several different companies, money is consistently ranked five or lower by most employees. So if money is not the best way to motivate your team, what is?

Employees' three most important issues according to employees are:

• Respect

• A sense of accomplishment

• Recognition

Question # 58

What is your biggest achievement?

Answer:-

Quality work to be is about doing work to the require or set standard, which is very important when it comes to warehouse operations.

Question # 59

How would you observe the level of motivation of your subordinates?

Answer:-

Choosing the right metrics and comparing productivity of everyone on daily basis is a good answer, doesn't matter in which company you apply for a supervisory role.

Question # 60

Have you got any questions?

Answer:-

This is your final opportunity to persuade the interviewer that you are the right candidate for the job. Now is not the time to ask questions about holidays, pay or pensions - all these things can be asked later when you get an offer of employment. Now is the time to ask about any reservations that the interviewer may have about your suitability for the role. You will then give yourself one last chance to persuade the interviewer that you are the right candidate for the job.

Example Thank you. I think we have covered everything. Before we finish the interview I would like to take the opportunity to ask if you have any reservations about my suitability for this role?

Question # 61

How would your friends describe you?

Answer:-

My friends would probably say that I'm extremely persistent - I've never been afraid to keep going back until I get what I want. When I worked as a program developer, recruiting keynote speakers for a major tech conference, I got one rejection after another - this was just the nature of the job. But I really wanted the big players - so I wouldn't take no for an answer. I kept going back to them every time there was a new company on board, or some new value proposition. Eventually, many of them actually said "yes" - the program turned out to be so great that we doubled our attendees from the year before. A lot of people might have given up after the first rejection, but it's just not in my nature. If I know something is possible, I have to keep trying until I get it.

Question # 62

What was the most important task you ever had?

Answer:-

There are two common answers to this question that do little to impress recruiters:

• 'I got a 2.1'

• 'I passed my driving test'

No matter how proud you are of these achievements, they don't say anything exciting about you. When you're going for a graduate job, having a degree is hardly going to make you stand out from the crowd and neither is having a driving licence, which is a requirement of many jobs.

Question # 63

Why should the we hire you as this position Regarding Erector Precast Concrete?

Answer:-

This is the part where you link your skills, experience, education and your personality to the job itself. This is why you need to be utterly familiar with the job description as well as the company culture. Remember though, it's best to back them up with actual examples of say, how you are a good team player.

Question # 64

What is your greatest weakness Regarding Erector Precast Concrete? What are you doing to improve it?

Answer:-

I believe my biggest weakness Regarding Erector Precast Concrete is wanting to help anyone I can help. What I mean is I am willing to take on task that are not my job. I want to learn all I can. However, that has helped me get promoted or even asked to help in times of need in other department. I have been know as the "go to person" when help is needed.

References.

1. Akash lanke and Dr. D. Venkateshwarlu (2016), 'Design, cost and time analysis of precast and RCC buildings' – International Research Journal of Engineering and technology, vol.3, No.6, pp. 343-350.
2. Dinesh kumar and Kathirvel P (2015)., 'Comparative study on prefabricated construction with cast in situ construction of residential building' – International Journal of Innovative Science Engineering and Technology, vol.2, No.4, pp.527-532.
3. Ninjal parekh and Bhupendra M Marvadi 'Comparative study of various construction techniques' – Journal of Information, knowledge and research in civil engineering, vol.3, No.2, pp.226-230.

4. Anderson, M., and Anderson, P. (2007). "Prefab prototypes Site-specific design for offsite construction." Princeton Architectural Press, New York, USA.
5. Tam, V.W., Tam, C.M., Zeng, S.X. and Ng, W.C., (2007). "Towards adoption of prefabrication in construction". Building and environment, 42(10), pp.3642-3654.
6. Jaillon, L., and Poon, C. S. (2008). "The evolution of prefabricated residential building system in Hong Kong: A review of the public & private sector." Automation in Construction, Vol. 18, pp. 239-248.
7. Workshop on Performance Evaluation of Housing Units IIT Kanpur, Extension Centre, Noida July 21-22, 2017
8. National Seminar on Green Affordable Housing June 13, 2014, New Delhi
9. Bendi, D., Arif, M., Sawhney, A. and Iyer, K.C. (2012). "Offsite Construction in India-An Exploratory Study". International Conference on Structural & Civil Engineering, pp. 64-67.
10. Bindurani, P., Meher Prasad A., and Sengupta, K. (2013). "Analysis of Precast Multi-storeyed Building-A Case Study." International Journal of Innovative Research in Science, Engineering & Technology, Vol. 2 (1), pp. 294-302.
11. Surya Prakash, P. (2013). "Pre-Cast Concrete Elements in Construction – Emerging Scenario in India." Precast Concrete Buildings-The Masterbuilder, pp. 128-134.
12. Purandare, S. (2013). "Precast in India-Latest Thinking". Precast Concrete in Residential & Commercial Buildings, www.builtconstructions.in
13. Brietbach D. (2014). "Indian Precast Construction-An Insight". Project Vendor Magazine (Weckenmann)
14. Bommi, S., Somaraju, K., Senou, K., and Barde, A. (2014). "Precast High-Rise Residential Projects in India: Design Implementation". The Masterbuilder, Oct-2014, pp. 106-114.
15. Dhoot, R., and Manocha, V. (2014). "Precast Concrete Construction: It's Challenges & Implementation for Mass Housing in India". REDECON, www.builtconstructions.in

16. Turai, V., and Waghmare, A. (2015). "A Study of Cost comparison of precast concrete vs. Cast-in-Place". International Journal on Recent & Innovation Trends in Computing & Communication, Vol. 3 (11), pp. 6235-6238.

THE END